D1520381

INFERTILE ENVIRONMENTS

Critical Global Health: Evidence, Efficacy, Ethnography
A series edited by Vincanne Adams and João Biehl

JANELLE LAMOREAUX

Infertile Environments

Epigenetic Toxicology and the Reproductive Health of Chinese Men

DUKE UNIVERSITY PRESS
Durham and London
2023

Designed by A. Mattson Gallagher

Project editor: Bird Williams

Typeset in Adobe Text Pro by Westchester Publishing Services

Library of Congress Cataloging-in-Publication Data
Names: Lamoreaux, Janelle, author.
Title: Infertile environments: epigenetic toxicology and the reproductive health of Chinese men / Janelle Lamoreaux.
Other titles: Critical global health.
Description: Durham : Duke University Press, 2022. | Series: Critical global health: evidence, efficacy, ethnography | Includes bibliographical references and index.
Identifiers: LCCN 2022027030 (print)
LCCN 2022027031 (ebook)
ISBN 9781478016700 (hardcover)
ISBN 9781478019336 (paperback)
ISBN 9781478023975 (ebook)
Subjects: LCSH: Medical anthropology—Research—China. | Reproductive toxicology—Research—China. | Genetic toxicology—Research—China. | Male reproductive health—Research—China. | Environmental health—Research—China. | Toxicology—Research—China. | Infertility, Male—Research—China. | Endocrine disrupting chemicals—Environmental aspects. | BISAC: SOCIAL SCIENCE / Anthropology / Cultural & Social | SCIENCE
General Classification: LCC GN296.5.C6 L366 2022 (print)
LCC GN296.5.C6 (ebook)
DDC 306.4/61072051—dc23/eng/20220803
LC record available at https://lccn.loc.gov/2022027030
LC ebook record available at https://lccn.loc.gov/2022027031

Cover art: Fluorescence in situ hybridization (FISH) of human sperm cells. Image by Aliza Amiel, from the paper "Interchromosomal Effect Leading to an Increase in Aneuploidy in Sperm Nuclei in a Man Heterozygous for Pericentric Inversion (inv 9) and C-heterochromatin," by Aliza Amiel, Federica Sardos-Albertini, Moshe D. Fejgin, Reuven Sharony, Roni Diukman, and Benjamin Bartoov. *Journal of Human Genetics* 46, no. 5 (April 2001): 245–50. Courtesy of Aliza Amiel.

In memory of G.G.S.

CONTENTS

PREFACE

Before returning to school for graduate studies, I worked at Cornell University's Center for Reproductive Medicine and Infertility (CRMI), a fertility treatment clinic in New York City's Upper East Side. As a patient coordinator at a clinic renowned for its high success rates and cutting-edge research, I helped patients from around the world navigate the difficult process of infertility treatment through reproductive technologies. Many of the patients I "coordinated"—a task that involved a combination of administrative processing, appointment scheduling, medical education, and impromptu counseling—had been unsuccessfully treated for infertility prior to their appointment at CRMI. During my two years in this position, between 2004 and 2006, I spoke with hundreds of individuals and couples undergoing treatment for infertility, many of whom had grappled with this diagnosis for some time.

Most of the couples I met were in their late thirties and early forties. Others were younger or older, from their early twenties to early fifties. Most seemed simultaneously frustrated by their infertility and determined to pursue treatment. Others were confused and dismayed, or excited and eager. Some of the patients I treated were single. Others were partnered in gay, lesbian, or queer relationships that remade heteronormative ideas of parenting and reproduction in creative ways. There were widows pursuing pregnancy with the sperm of their deceased partners, and divorcées making time lines work with the help of best friends. There were former cancer patients pursuing in vitro fertilization (IVF) who had frozen their eggs or sperm before radiation or chemotherapy. There were transnational couples undergoing intrauterine insemination (IUI) because they were never together in the right place at the right time. Some patients were middle-class or low-income and seeking services through a grant program sponsored by the state of New York. But most of the people I helped treat were wealthy, heterosexual couples of a variety of ethnicities, nationalities, religious affiliations, and professions who had been trying to conceive for six months or longer and had not "achieved pregnancy" (as they say, as if such acts are a personal accomplishment) or whose pregnancy or pregnancies "had not resulted in a live birth" (in other words, had ended in miscarriage or stillbirth).

During my interactions with patients, they often spoke about how difficult it was when their predicted path to parenthood had been derailed.[1] Women diagnosed with "advanced maternal age" often expressed the feeling that this was their last chance to be parents. They didn't know they wanted to have children so much until it was too late or had only now met someone with whom they wanted to have kids.[2] In and outside of the clinic, questions of age permeated discussions of infertility, especially "female-factor infertility." Sex-specific structural abnormalities such as polyps and endometriosis also came up, but less frequently. In my discussions with men diagnosed with "male-factor infertility," age was not often brought up. But they too expressed anxiety, often over the phone, during calls that took place after our in-person meetings. I answered nervous questions about the semen-collection process and repeated the details of semen-analysis procedures so many times that I had a script for nearly every nervous query. Other men and women faced the difficult challenge of being diagnosed with "unexplained infertility" and lamented the lack of clarity that such an enigmatic label conferred.

Statistically speaking, unexplained infertility is said to affect approximately 20 percent of those seeking infertility services, with the 80 percent of remaining cases shared equally between male factor and female factor. To me these unexplained diagnoses, as well as the inexplicable dimensions of common biological explanations for infertility, were the most difficult to discuss. In situations where the cause of infertility was more obvious—for instance, in cases of advanced maternal age, structural obstruction, or azoospermia (the absence of sperm)—there seemed to be physical explanations for difficulty conceiving. But with unexplained infertility, answers were less clear. Low sperm counts, irregular periods, polyps or fibroids, low sperm quality, anti-sperm antibodies—these were conditions often discovered during diagnostic procedures. However, they weren't really causal factors, more like inhibiting symptoms of a bigger problem or broader issue, poorly understood. I increasingly found myself thinking about what was making sperm decline, polyps grow, periods stagger, and cervical mucus become "hostile" to sperm.[3] What were the root causes of infertility, and why weren't they being discussed? Physicians often prescribed reproductive technologies to patients in such unexplainable circumstances; as reproduction scholar Sarah Franklin writes, "Into the breach of explanation is inserted a technological enablement" (1997, 322). A lack of explanation first enabled costly diagnostic procedures, then IUI, then IVF, then IVF with intracytoplasmic sperm injection (ICSI) or pre-implantation genetic diagnosis (PGD) . . .

My desire for a more generous rendering of infertility's cause increased as I was repeatedly confronted with patients' questions of why. Why can't I get pregnant? Why is it so easy for others my age to conceive? Why am I infertile? Such questions take on heightened meaning at a moment when individuals are often blamed for health problems. Questions about why a specific person is infertile can quickly turn into answers that stress individual responsibility for reproductive health. For instance, patients often wondered if they had done something to cause their infertility or were doing something wrong in their efforts to conceive. They looked for suggestions for things they could do, or take, or eat, or abstain from to make themselves fertile. Not only do such pursuits show how determined many people are to meet personal, social, cultural, and familial expectations of biological relatedness.[4] They also show how much people internalize the idea that pregnancy is an accomplishment (Becker 2000). This is especially the case for women, since infertility has historically been conceptualized as a women's problem, and women's bodies continue to take center stage during infertility treatments (Thompson 2005).

While working at the clinic I witnessed the gendered organization of fertility treatments. I watched thousands of women going through an endeavor that demanded weeks, months, and even years of fertility treatment, and the corresponding way that men's role in the process is sidelined.[5] This occurs partially because of physiology—particularly differences in the accessibility of eggs and sperm—and the need to prepare and monitor pregnant people's bodies for surgical procedures and their aftermath. But it also occurs partially because of the way that treatment practices, procedures, and prerequisites are organized (Sandelowski and De Lacey 2002; Thompson 2005). For example, even when couples that I helped treat faced a known male-factor issue, extensive diagnostic tests were often required of women. These included not only simple blood draws but also more elaborate and expensive procedures such as hysterosalpingograms and laparoscopies. The most common treatment for male-factor diagnosis, ICSI, requires that women undergo IVF, which involves weeks of hormonal injections, multiple transvaginal ultrasounds, and the surgical retrieval of eggs as well as the implantation of embryo(s). The necessary extent of women's role in infertility treatments has been exacerbated by the gendered focus of reproductive science and medicine.

Many scholars of reproduction have more thoroughly explored these gendered aspects of reproductive technologies in and outside of fertility clinics. When I began attending graduate school at the New School for Social Research in 2005 while still working at the clinic, I began reading this social science of reproduction and merging my interest in the gendered experiences of infertility with the causal questions that patients and I had asked. I quickly embraced a feminist critique of the "biological clock" that recognized the structural reasons for people delaying pregnancy and attributed rising infertility rates to the gendered organization of social and economic life (Friese, Becker, and Nachtigall 2006). Still, such reflections on age and work did not seem to capture the concerns of people whose diagnoses fell within the 20 percent of unexplained infertility or even the 40 percent of patients diagnosed with male-factor infertility (which at the time was not commonly linked to age).[6] Although those I had worked with at CRMI researched the causal factors of infertility from the standpoint of individual bodies, I wondered if others were thinking through conditions of life outside the individualized body that might lead to infertility.

Once I stepped outside the clinical setting, both professionally and conceptually, I found that a broader perspective on the potential causes of infertility did exist. In the research of some endocrinologists, andrologists, and even toxicologists, infertility was viewed almost as a side effect—an aftershock of industrialization, institutional policies, inequitable legacies, and pollution. This was especially true in the case of men's infertility, which had a more outward-facing orientation, while research on women's infertility seemed to still be more tethered to the body (Martin 1987). Studies of sperm decline emphasized not only individual behaviors, genes, and physical characteristics but also the effects of occupational settings, pollutants, food, and household products on male developmental and reproductive health. In my reading, it seemed that such research almost characterized male infertility as a proxy diagnosis for a world order struggling to reproduce itself.

This book is an effort to further articulate these high stakes by shifting the study of infertility beyond the individual toward "the environment." Today, *environment* can mean many things, and its usage is often problematically reductive or muddled (Keller 2002). But what happens to the idea of infertility when it is stretched to environmental scales? "The woman in the body" might be situated within multiple contingent relations, infrastructures, exposures, and imaginaries (Petchesky 1987; Martin 1987). Reproduction might be reconsidered as an act that does not "end at our bodies" (Murphy 2013). Reproductive technologies might be reconceptualized to include those techno-scientific artifacts and arrangements of everyday life that exist outside biomedical clinical settings (Haraway 1997b). And all politics might be understood as reproductive politics (Briggs 2018; Ginsburg and Rapp 1991).

To study infertility as a condition that comes about in specific and dynamic political, economic, social, and chemical contexts is to expand not only infertility's etiology but also its applicability. Understanding infertility as an environmental issue moves beyond the individual, beyond the partnership or the choice, beyond clinical diagnosis of advanced maternal age and poor sperm motility, to a wider diagnostic lens. Of course, such an approach is not immune to the gendered stereotypes that permeate infertility treatment and research. Histories of individualism and biological determinism are also conscripted into ideas of environmental health. Consequently, an environmental approach to infertility cannot replace the idea of infertility as individualized reproductive failure. But such a perspective might encourage people to regard infertility as more than an individual's inability to reproduce another individual. The reproductive toxicology I discuss in the remainder of this book is an imperfect tool through which a reworking of infertility might continue.

ACKNOWLEDGMENTS

Thank you to everyone in Beijing, Shanghai, and especially Nanjing who, more than ten years ago now, enabled this research. You befriended me and invited me to events, dinners, hikes, site visits, interviews, and karaoke. Thank you especially to those who worked at the DeTox Lab during this time for enduring my many questions during formal and informal interviews and allowing me to shadow your work and accompany you inside and outside research settings. Thank you to the many other faculty, students, and researchers in China who hosted and helped me while I was developing this project and as I conducted research.

Thank you to my postdoctoral supervisor, Sarah Franklin, for creating an amazing community of reproductive scholars and inviting me to be a part of it; your feedback, support, and scholarship were crucial to writing this book. Thank you to my doctoral advisor, Cori Hayden, whose patience, advice, suggestions, support, and thoughtful questions during dog-walks around North Oakland saw me through my PhD at Berkeley. Thank you to Vincanne Adams, Lawrence Cohen, and Yeh Wen-Hsin for everything you did as dissertation committee members and continue to do as mentors to support me intellectually and professionally. Thank you to my professors at the New School, especially Hugh Raffles, who supported me in my master's education and encouraged me to pursue a PhD. Thanks also to the SO/AN faculty at Lewis & Clark College for growing my interest in social theory and anthropology, and helping me find an academic home in the social sciences.

Thank you to those who gave me feedback on this work in progress. This includes participants in an early manuscript workshop and in the ChinaReproTech conference, including Sarah Franklin, Katharine Dow, Ayo Wahlberg, Joy Zhang and Nancy Chen. Thank you to the many editors and anonymous reviewers who offered feedback on pieces of this text on their way to publication in various venues, including Cymene Howe, Ruth Rogaski, Emily Yates-Doerr, Christine Labuski, Andrea Ford, Heather Paxson, Iris Borowy, Alex Nading, and Sahra Gibbon. Thank you to the many individuals and institutional groups who invited me to speak over the years, and the conversations that followed such engagements and transformed this book, especially Megan Carney and the University of Arizona Center for Regional Food Studies Seminar participants; Matthew Kohrman and faculty and graduate students at Stanford Anthropology; Alex Nading and participants in the Embodied Being, Environing World Symposium; Zoe Wool and the faculty, postdocs, and students at the Chao Center for Asian Studies and Department of Anthropology at Rice University; and members of Bharat Venkat's 2020 graduate seminar "The Body in Milieu." Conversations and workshops with Rene Almeling, Dwai Banerjee, David Bond, Joe DeStefano, Lyle Fearnley, Brie Gettleson, Anna Jabloner, Gregor Sokol, Hentyle Yapp, Jerry Zee, and many others were crucial to moving this project along.

Thank you to the many scholars who have written about and taught me about reproduction and feminist science studies more generally. Through courses, workshops, reviewer comments, and feedback on presentations, you build a supportive group that helps so many of us make our way through the academy and the world. I feel lucky to share in such community and to have taken courses with so many amazing academic women: Rayna Rapp, Adriana Petryna, Adele Clarke, Donna J. Haraway, Charis Thompson, and others.

Thank you to my colleagues at the University of Arizona who have supported me as I wrestled this manuscript into being, especially Megan Carney, Rayshma Pereira, Eric Plemons, Catherine Lehman, Emma Blake, Lars Fogelin, Ivy Pike, Hai Ren, Brian Silverstein, Stefanie Graeter, Tom Sheridan, Jen Croissant, and Diane Austin. Thank you to my research assistants, M. Bailey Stephenson and Siwei Wu, as well as Emma Bunkley, Brittany Franck, and Rinku Ashok Kumar, who helped me at various stages of this project.

Thank you to Ken Wissoker, series editors Vincanne Adams and João Biehl, and Joshua Gutterman Tranen for their support of this book, anonymous reviewers for their helpful suggestions for improving the text, and all the editorial and production staff at Duke University Press for bringing the book into its final form.

Thank you to those of you who have stuck with me through the ups and downs of scholarly and personal life. I am especially thankful to Amber Benezra, Katie Dow, Marieke Winchell, Ruth Goldstein, Karen Jent, and Bharat Venkat. Thanks also to my friend and former coworker Andrea Felton for always entertaining my queries about reproductive technologies, even long after I left the Center for Reproductive Medicine. In many ways our conversations sparked the original version of this project, which has now grown into something nearly unrecognizable, much like ourselves.

I am grateful for a kin network that believes in me, supports me, and accepts me quirks and all. This includes Linda, Brendon, Michael and Joan, Paul, Jean, Aidan, Jose Pablo, Iona and Susan, Garry and Mary, Sierra and Emma, and the extended Riffle family—especially Jason and Eli. This book is dedicated to my grandfather, recently passed, whose intergenerational effects are enduring.

This research was supported by the Wenner Gren Foundation, the National Science Foundation, the Social Science Research Council, the University of California Center for Chinese Studies, and the University of California Center for Science and Technology Studies, as well as the University of Arizona Social and Behavioral Sciences Research Institute.

Introduction

In 2013 the meteorological association of the People's Republic of China released a green paper on climate change that highlighted both the economic and health effects of pollution. In a description of how urbanization in China had increased the country's carbon footprint, the report briefly mentioned the potential negative influence of pollution on human reproductive health (Chao 2013). Media coverage and commentary quickly focused in on this provocative connection. Journalists began interviewing fertility experts who described a "sperm bank emergency" (*jingzi ku gaoji*) even more exaggerated than sperm shortages that had come before. Since 2002, multiple regional sperm bank emergencies had been declared in China as local sperm banks reported that the quality of donor sperm was in decline (Wahlberg 2018b). Initial speculation about the cause of this decline focused on lifestyle factors, diet, and stress levels. The rise in standards of living during the past thirty years of reform and opening (*gaige kaifeng*) policies had dramatically changed everyday habits, occupations, and living arrangements in China, bringing many people out of poverty. But the potential drawbacks to such economic advancement were also becoming apparent—for instance, through discussions of a decline in men's sexual and reproductive health (E. Y. Zhang 2015).

Ten years after these first regional shortages had been publicized, sperm decline appeared to be both lasting and widespread.

Urological experts attributed the increasingly chronic nature of such "emergencies" to several factors. These included high institutional semen-quality standards, sexually transmitted diseases among potential donors, policy limitations on donation usage, and hesitations to donate biomaterial because of "cultural hurdles" (Wahlberg 2018b, 101; Ping et al. 2011).[1] But unlike past emergencies, men's reproductive health experts were also now more forthrightly drawing connections between the decline of sperm and the rise of industrial pollution. Sperm bank emergencies had gone from regional concerns about the effects of lifestyle changes among China's young men to a topic of national conversation about the pollution of China's environment. In the straightforward words of one sperm bank coordinator, quoted in the *Shanghai Morning Post*, "If the environment is bad, sperm become ugly" (L. Chen 2013).

Many people reacted to this report of ugly sperm in social and print media outlets. Some newspaper commentators made practical suggestions for individuals in light of the lack of immediate solutions to widespread environmental problems. They recommended avoiding smog by staying indoors and eating detoxifying foods to preserve fertility (C. Zhang 2013). But many others interpreted sperm decline as more than an individualized problem, and instead as a broader issue with national and intergenerational dimensions. Hundreds of users of the popular social media platform Weibo responded to the news story. Many made serious jokes, wondering if pollution was perhaps the latest version of China's notorious birth-planning policies. Others stressed the new meaning that old sayings seemed to take on in an era of environmental pollution; as one Weibo user wrote, "Before I didn't understand the saying 'beautiful mountain, beautiful water, and beautiful people.' Now I understand" (kingarthurzj_9006 2013). Still others emphasized the intergenerational stakes of sperm decline in a "bad" environment: "The cost of society's development is sacrificing the next generation. Sad!" (Jinhuozaifendou 2013). Echoing such intergenerational sentiment, the reporter for the *Shanghai Morning Post* who originally reported on the rise of ugly sperm wrote that "in the view of fertility experts, taking care of the earth equals taking care of ourselves and of the next generation" (L. Chen 2013).

Such reactions to the story of ugly sperm showcase an argument central to this book and at the heart of much feminist analysis of reproductive substances: when people anxiously discuss the decline of reproductive potential, they are talking about much more than a threat to individual fertility. They are also talking about a threat to the reproduction of social, national, and economic order. Such an argument is not just a social scientific talking point; it is an interpretation increasingly made by people around the world who are concerned about the intergenerational repercussions of increasing toxic exposures. How toxicity—in material and immaterial forms—influences the ability of people and other beings to reproduce physically and culturally is increasingly articulated as an urgent question by and for many (Dow 2016; Hoover 2017).

As a cultural anthropologist conducting research in Nanjing between 2008 and 2011, I found that experts, activists, and scholars often connected concerns about reproductive health with reflections on economic, political, social, and environmental change. More so than asking about individuals' responsibility for conditions such as infertility, people were talking about their reproductive health as intrinsically entangled with multiple environments and factors. This book explores how and why reflections on the causal factors of infertility are being reimagined and redefined at a moment of growing attention to toxic exposures and pollution. Why is men's infertility, in particular, and reproductive and developmental health, more generally, such an important lens through which people understand the imbalance of their relationship to one another and to "the environment"? What does the environment mean to those researching and otherwise reflecting on its relationship to reproduction and development?

I approach these questions through a focus on epigenetic toxicology. Today, *toxicology* is defined as the study of the potential harmful effects of "chemicals, substances, or situations" on humans and animals ("Toxicology" 2019). Epigenetic research, frequently referred to simply as "epigenetics," is typically thought of as the study of modifications to genes that affect gene expression without changing the sequence of DNA. Since the turn of the twenty-first century, epigenetic toxicology has increasingly drawn attention to the way that potentially harmful environmental exposures influence DNA expression. But how the environment gets defined in epigenetic research is more complicated than it first may seem. Through a study of toxicologists based in Nanjing who practice epigenetic research, in this book I ask: If "the environment" is to blame for the decline of men's reproductive health, then what kind of environment is it? In part because of epigenetic research, answers to this question have both proliferated and changed.

What Is the (Chinese) Environment?

The Chinese word most often translated into environment is *huanjing*. In modern Chinese *huanjing* frequently describes a natural setting, often in need of protection from humans by humans. But, as with the English word *environment*, naturalness is only one of many meanings and connotations of *huanjing*. Today this term can be found in various conversational venues, government campaigns, and business arenas. Besides the natural environment (*ziran huanjing*), there are investment environments (*touzi huanjing*) and working environments (*gongzuo huanjing*), recreational environments (*yule huanjing*), family environments (*jiating huanjing*), and social environments (*shihui huanjing*) (Hoffman 2006). These multiple environments do more than simply give terminology to a growing set of preexisting entities. Environments are brought into being through the practices that make them knowable objects. Translated from Chinese to English, *huán* means ring or circle, and *jìng* means condition or circumstance. The environment is a circumscribed set of circumstances; enclosure itself makes the environment. Historian Chia-Ju Chang similarly breaks down the individual characters of the word *huanjing*, arguing that in its premodern usage the term was a means of nationalist place making. She calls this place-making practice "environing" (Chang 2019). By using *huanjing* as a verb instead of a noun, Chang dislodges the term's natural and stable connotations. Instead, environments in the making are emphasized.

This book takes inspiration from such interpretations, and from a long history of anthropological thinking that similarly attends to practice, including Judith Farquhar's research on infertility and Chinese medicine (*zhongyi*) that interprets objects as processes (Farquhar 1991).[2] By focusing on epigenetics in practice and on the environment as a process, this book shows how epigenetic environments are brought into being during research. It also shows how environments that are materialized through research practices reverberate with environmental concerns that take place outside of research venues. Environments are not only multiple, enacted by persons and through technologies in various ways (Mol 2002); they also come into being in relation with other environmental forms that simultaneously exist at multiple scales and in numerous domains.

Today, protecting the environment is a large part of the official Chinese Communist Party (CCP) platform and is often a part of people's everyday reflections on the state of China's air, water, land, and food.[3] Environmental protection has been declared a national priority. It is also an international political strategy that foregrounds China's climate-change mitigation and sustainability efforts. Environmental consciousness has become a marker of both modernity and cosmopolitanism, and it is supported by a government that expresses deep concern about the state of the environment nationally and globally (Hubbert 2015; J. Y. Zhang and Barr 2013). The risk of environmental pollution to reproductive health has now been raised by many as a factor of personal and familial concern and one that potentially undermines the ability of people in China to have and raise healthy children (J. Li 2020; Wahlberg 2018a). But such widespread formal and informal acknowledgment of the likely connection between environmental and reproductive health was less present during the time of my fieldwork.

This book is primarily based on fieldwork conducted in between 2008 and 2011, at a time when the environment was not as prominent of a concern among those I met in China as it is today, more than ten years later. This was before Premier Li Keqiang's 2014 declaration of a war against pollution. It was before the viral circulation of *Under the Dome* (*QiongDing Zhi Xia*), a TED-talk-style documentary made by Chai Jing, a former China Central Television employee, that highlights the link between environmental pollution and health, in particular the health of her young daughter, who was diagnosed with a heart defect in utero. This was before the series of "airpocalypses" that descended upon Beijing and other locations; before the mass adoption of face masks and home air-filtration systems—what anthropologist Matthew Kohrman (2020) calls "filtered life"; before actress Zhang Ziyi announced she was leaving the country out of fear for her young daughter's developing lungs, and before some reacted to this announcement by pointing out that her ability to walk away from pollution was a privilege.

But 2011 was also a time when the quantity of toxic exposures faced by those living in many parts of China was clearly growing. Protests against specific commercial enterprises and development projects, often surrounding the waste generated by industrial and energy projects, had erupted throughout China and were growing in number by the year (Steinhardt and Wu 2016; B. Wang 2019; A. Zhang 2020). Environmental litigation had emerged as a "politically touchy, but not taboo" means of seeking compensation for pollution events (Stern 2013, 2). Toxic chemical exposure was being researched by a growing number of regional and international nongovernmental organizations (NGOs) and being covered by an increasing number of media outlets (J. Y. Zhang and Barr 2013). For many, the environmental protection (*huanjing baohu*) and ecological civilization (*shengtai wenmin*) stressed by the government felt more like an act of transnational diplomacy than an effort to care for the health and well-being of present and future citizens.[4]

In such contradictory conditions, toxicity becomes more than simply a measure of capacity to bring about harmful effects. Toxicity increasingly becomes a material *and* existential concern through which people struggle to make sense of political-economic policies and distributed social hierarchies, as well as their consequences. In the university-rich city of Nanjing, where I conducted fieldwork, graduate students and professors from disciplines as varied as environmental science, medicine, philosophy, and toxicology were thinking through what this burgeoning attention to toxicity and the environment meant for their country, their region, and their lives. My research focused on a small group of toxicologists that I refer to as the DeTox Lab.[5] At the DeTox Lab, research on the reproductive and developmental influences of environmental exposures was the vehicle through which such thinking about toxicity occurred. In the lab's research, the environment is circumscribed at many scales and comes to mean many things. It is the food, air, and water that is taken into people's bodies, as well as the specific chemicals in these substances. It is the factory, the city of Nanjing, the Yangtze River Delta region, and the nation of China. It is a person, a mother, and a body—variably predisposed to influence along gendered and racialized lines. The environment is materialized through their epigenetic research as all these things and more.

Epigenetic Im/Possibilities

Epigenetic research hypothesizes that a wide array of things previously thought to have no impact on genes are now understood to modify gene expression (Landecker and Panofsky 2013). Conditions such as poverty, lifestyle factors such as diet, or events such as famine or trauma are now referred to as exposures or environments that are thought to have epigenetic effects. The way we live our lives, the environments within and around us, and the things we are exposed to are thought to change the expression of DNA, even though they do not change DNA themselves. In addition, epigenetic research often investigates the potential intergenerational effects of these modifications through animal experiments and birth-cohort research on intergenerational inheritance.

Despite contemporary agreement on this general definition of *epigenetics,* the term is actually quite difficult to pin down. C. H. Waddington's 1942 conceptualization of epigenetics focused on developmental effects. Today the term is used by an increasing number of research groups and disciplines to describe a wide variety of research approaches. According to entomologist

FIGURE I.1 This comic was shown during a presentation I attended at an epigenetics conference at the University of Cambridge in 2014. It both questions and reproduces the explanatory power of epigenetics.

Carrie Deans and biologist Keith A. Maggert, such varied usage has led to a lack in consolidation of epigenetic meaning among natural scientists. In an article titled "What Do You Mean, 'Epigenetic'?" they argue that epigenetics has become "a catchall for puzzling genetic phenomena" (2015, 889). Such sentiment seems prevalent among researchers who use epigenetic research techniques, as depicted in a comic that was shown at the end of a presentation at an interdisciplinary epigenetics conference I attended while at the University of Cambridge, which brought together biologists researching across many species and specializations (see figure I.1). The image both questions and reproduces the explanatory power of epigenetics.

Making a similar point from a social scientific perspective, anthropologist Marilyn Strathern (1991) has called epigenetics a "biologist's catchall." Strathern was among the first social scientists to note the turn toward epigenetics in the reproductive sciences, describing the term's meaning in a 1991 publication as a focus on "everything else besides the gene." This shift placed epigenetics' potential area of research in a different order, in Strathern's words "imagined, hypothetically and thus abstractly, as infinite." Nevertheless, epigenetics, she argued, was concretized through a concept of the environment that could be made to stand for diverse contexts, "as uterus or as trees and mountains" (1991, 586). Since the time of Strathern's observation, the meaning of both epigenetics and the environment continues to simultaneously take on greater ambiguity and more concreteness. Today the infinite space "beyond the gene" continues to be concretized as environments that are made to stand for multiple contexts and things, and epigenetics itself is made to stand for many modes of investigating environmental-influence on gene expression.

Despite or perhaps because of its definitional ambiguity, epigenetics has been touted by many as a revolutionary way of thinking about inheritance (see Carey 2012). Popular accounts in books and magazines often depict epigenetics as a departure from what is often regarded as DNA's iconic place in modern Western consciousness (Franklin 1988, 95; Nelkin and Lindee 2004). Epigenetics has been interpreted as a potential corrective to the gene-centric view of health, disease, and even fate. By moving to that which lies beyond the gene, epigenetics potentially diminishes the power of the gene, showing how biology in general, and in parsed biological units such as sperm, is shaped by what stands beyond it.

However, many remain skeptical of epigenetic research practices and the revolutionary label affixed to such pursuits. Social scientists have shown that so-called post-genomic approaches, which claim to go beyond the gene in their studies of genetic expression and inheritance, often rely upon and conscript a genetic approach (Landecker 2016; Gibbon et al. 2018). In the past, oversimplified ideas of DNA as "the code of life" led to genetic determinism; now, oversimplified ideas of the power of environmental factors in determining future health have led to "epigenetic determinism" (Waggoner and Uller 2015). A fixing of sociocultural factors as stagnant and bounded environments occurs in epigenetic research in ways that sometimes perpetuate gendered and racialized stereotypes (Kuzawa and Sweet 2009; Saldaña-Tejeda 2018; Saldaña-Tejeda and Wade 2019; Valdez 2021), or obfuscates complex structural realities through a reductive vision of environmental factors or "social determinants of health" (Yates-Doerr 2020). In this sense, epigenetics' connection to the essentializing force of genetics is again quite strong. Such persistent essentializing has led historians of science to describe epigenetics as more of a recycling and coexistence with past ways of thinking about development and inheritance rather than a revolutionary paradigm shift (Meloni and Testa 2014; Peterson 2016).

Still, epigenetic thinking and research offer a depiction of biology that partially overlaps with scholarly work that has historicized and complicated the category of "life itself" (Franklin 2000). Epigenetic research stresses that biology is not something that is given but that is constantly being made and remade, be it through environmental exposures (Fortun 2011) or technological interventions (Franklin 2013a; Hayden 1995; Thompson 2005). Through epigenetics, scientists are—for instance—considering that environments influence bodies and health in a way that is more reminiscent of multiple "alternatives" to Western biomedicine. Moreover, epigenetic thinking also aligns with many Indigenous perspectives on the entanglement of human and nonhuman ontologies (Warin, Kowal, and Meloni 2020). If, then, epigenetics is recreating ontologies (Lock and Palsson 2016), it is doing so through relational vocabularies that have long existed in many communities, languages, and traditions not frequently privileged by biomedicine—an ontological heritage that often elides STS scholars (Todd 2016).

Epigenetic Lineages

A history of epigenetics centered on Europe and the United States often highlights the multitude of approaches to genetic thinking in the twentieth century, which congealed into a dominant theory of genetics and DNA by the century's midpoint.[6] This history often begins by pointing to the overlap of current epigenetic theories with Lamarckian ideas about the inheritance of acquired characteristics (Jablonka and Lamb 2006; Rapp 2005). (Neo-) Lamarckianism went out of fashion in the early twentieth century with the 1900 rediscovery of Gregor Mendel's rules of inheritance—which stated that inheritance works through discrete units passed from parent to offspring. According to historian of science Evelyn Fox Keller, despite Mendelianism's strong influence on the science of that time, the first four decades of the twentieth century continued to be riddled with questions about what actually constitutes a gene (Keller 2002). However, with the 1943 identification of "DNA as the carrier of biological specificity," then the 1953 announcement that "genes are real molecules" made up of deoxyribonucleic acid (DNA), consensus began building around the constitution of the gene. "Thus, by midcentury," Keller writes, "all remaining doubts about the material reality of the gene were dispelled and the way was cleared for the gene to become the foundational concept capable of unifying all of biology" (2002, 3).

But one could also trace a different epigenetic lineage that includes an alternative understanding of what constitutes the gene, as well as a different understanding of genetics' centrality to twentieth-century ideas of inheritance. During China's Republican Era (1912–1949), neo-Lamarckian and Mendelian genetics were not understood as mutually exclusive. Whereas growing exploration into the science of heredity (*yichuanxue*) was informed by the return of Chinese geneticists and biologists who had studied in Euro-America, those practicing the sciences of heredity merged neo-Lamarckian and Mendelian genetics, stressing the interdependence of nature with nurture and connecting their science to a burgeoning commitment to strengthen the Chinese "race" (*minzu*) and nation (Dikötter 1998, 118). This changed after the Communist Party took formal control in 1949, when Mendelian genetics was denounced as bourgeois science and affiliated with the eugenic campaigns of Adolf Hitler and the hegemony of Western science. The "Morgan school of genetics" or "Morganism-Mendelism" was criticized for focusing too tightly on chromosomes as hereditary material and was banned by the CCP for its imperialistic, idealistic interpretation of generational continuity (Jiang 2017).

Following a policy of "learning from Russia," the party instead adopted Lysenkoism, a theory of heredity based on the work of Trofim Lysenko. Lysenko was an agronomist and biologist who emphasized the "relation of an organism of a given nature to its environmental conditions" (Lysenko 2001 [1951], 7). Like neo-Lamarckianism, Lysenkoism stressed the malleability of inheritance and the responsiveness of organisms to their environments. Such an approach, credited by Lysenko to Russian plant biologist Ivan Vladimirovich Michurin, became known as "Michurnist biology" and was sanctioned by the CCP. Michurnist biology's theory of hereditary adaptations to environmental changes fit with socialist dialectical materialism at the heart of CCP doctrine (Schneider 1989). Lysenkoism folded the history of the Chinese people into the history of the material world that surrounded them, arguing that plants and potentially human bodies would carry histories within them (with the understanding, of course, that humans make history).

Geneticists in China were not allowed to openly teach the Morgan school or conduct Morgan-Mendelian genetic research from 1949 until 1956, when transformations in the Soviet Union's political leadership and heightened utilitarian concerns led CCP leader Mao Zedong to readjust national policies (P. Li 1988). This led to the Hundred Flowers movement, where citizens were encouraged to openly express their stances on various issues, including science, during organized events.[7] As a result, some research with non-Michurnist leanings became sanctioned.[8] But work on both sides of the Michurnism-Mendelism approach came to a halt during the Great Leap Forward (1958–62), a period when a CCP-led campaign to move China from an agricultural to an industrial economy resulted in one of the deadliest famines in history, referred to as The Great Famine or "Three Difficult Years" (*sannian kunnan shiqi*) (E. Zhang, Kleinman, and Tu 2010).[9] Scientific recovery from this devastating famine (1959–61) was brief. The Cultural Revolution began in 1966 and saw the closing of almost all laboratories. Many scientists, regardless of their theoretical commitments, were criticized for following foreigners and losing touch with reality, and eventually sent to the countryside for "reform through manual labor" (P. Li 1988). Self-reliant (*tu*) science was emphasized and perceived as superior to foreign (*yang*) science, reflecting a binary that mapped onto comparisons of Chinese (*zhong*) and Western (*xi*) (Fu 2017).

It wasn't until after Mao's death and subsequent transfers of power that universities and laboratories resumed regular activity. After Deng Xiaoping took power in 1976, the national government reoriented toward economic development through reform and opening policies as well as the "four modernizations" campaign, which included a focus on science and technology. Competitive state funding for scientific research increased, as did the possibility of connecting technological development to commercialization opportunities inside and outside of China. Still, laboratory conditions remained poor through the eighties, even as a national scientific infrastructure was reestablished (Jiang 2015). But by the 1990s, China was developing a place in the increasing internationalization of science and would make major contributions to the Human Genome Project (Z. Chen and Zhao 2009). By the turn of the century, genomic scientific infrastructure, funding, and contributions through international publications and collaborations were growing faster in China than anywhere else in the world (Greenhalgh and Zhang 2020).[10] At this same moment of growth in China's genomic sciences, theories of the centrality of the gene to inheritance, health, and identity were increasingly being questioned by researchers in Europe and the United States.

In the aftermath of the Human Genome Project and its failure to identify meaningful genetic diversity, the limits of DNA's predictive power resulted in renewed interest in theories of gene-environment interaction (Shostak 2013). But it would be a mistake to think of such global rise in a gene-environment interaction approach as the single inspiration for epigenetic research in China, including the DeTox Lab's epigenetic toxicology. Many Chinese scientists had been required to train in both biology and Chinese medicine,

Mendelian genetics and dialectic philosophies—a dual emphasis established in the mid-1950s as part of the CCP's commitment to integrating Western and Chinese medicine, while nevertheless emphasizing the superiority and transformative potential of Chinese medicine (*zhongyi*) (Fu 2017, 132). Many grew up understanding food as medicine, or as operating through a principle of health and medical practice that viewed a correlation between inner states and outer conditions. Many lived through intense moments of socioeconomic and political transformation that had resulted in rapid changes to health, wealth, and everyday living, including changes to medical care, food, and education. Such intellectual and material surroundings shape, though certainly do not determine, how those researchers who I studied make sense of the connections between interior and exterior, exposure and effect, gene and environment. The DeTox Lab built from an idea that genes or DNA, even though important, were never immune to the environment.

Situating Epigenetic Research as Method

As mentioned, social scientific studies have shown that while the idea of epigenetics has the potential to reimagine the limits of the biological, such research can also do harm by reifying gender and racial stereotypes. While this is an important point that I continue to stress in this book, sometimes these same social scientific critiques of epigenetics themselves lack situatedness—a sense of the historical, political, cultural, socioeconomic, and other factors that influence how knowledge comes into being through particular people at certain times in specific places (Haraway 1988). At times, social scientists describe epigenetic research and discourse as if it exists outside the situations in which it is practiced. This obfuscation perpetuates the assumption that the values, ontologies, and imaginaries of the places in which social scientists most frequently study epigenetics (i.e., the United States and Europe) are the universal default, and that social scientists' critiques of epigenetics in general apply to epigenetic research practices and everyday imaginaries of epigenetics everywhere. Certainly, epigenetic research practices work in and through transnational assemblages of scientific infrastructure, research trends, and scientific languages and vocabularies (Ong and Collier 2004). But they, like other scientific practices, are also influenced by socioeconomic inequities (Tousignant 2018), national and regional funding mechanisms (J. Y. Zhang 2012), and ethical expectations and configurations (Ong and Chen 2010). Situated expressions of sexism, racism, nationalism, and individualism also influence the assumptions and stereotypes embedded in scientific research. In short, the limitations and possibilities of epigenetic research express and produce contingent political, socioeconomic, and historical conditions as well as values, norms, and imaginaries of kinship, gender, race, and inheritance.

This book is methodologically informed by primarily two types of research which both contribute to situating epigenetics. First, I conducted participant observation and approximately twenty-five interviews during fieldwork in China. I formulated this research project between 2008 and 2010, spending time in Beijing, Shanghai, Nanjing, and Chongqing. I eventually settled on Nanjing as the location of my long-term fieldwork, conducted in 2011, primarily because of the location of the Detox Lab but also because of the city's rich history of scientific research and transnational knowledge production. As the former capital of China, Nanjing has long been an educational center and today contains more than forty universities. During fieldwork, I regularly spent time with faculty and graduate students from multiple universities in the city as well as Nanjing residents unaffiliated with these educational institutions. I met these residents—who were employed in various sectors, from energy to tourism—through engaging in local activities and through connections I had made during earlier preliminary research.

In 2011, I lived in downtown Nanjing, where I rented a small apartment in a building occupied primarily by Chinese families. This building was within walking distance of the Nanjing Institute of Medicine and Science (NIMS), where the DeTox Lab is based. At NIMS, most of my time was spent with members of the DeTox Lab. During my days at the lab I observed studies in the laboratory and through such observation learned more about how environments were brought into being during epigenetic research practices. I interviewed DeTox Lab members as well as affiliated physicians and scholars, and I also joined in meals and leisure activities. I sometimes assisted with the work of the scholars I observed and interviewed, copyediting English-language articles before resubmission and English-language presentation slides. I also attended presentations in the Toxicology Department's seminar series, toured various laboratories, and met with faculty and graduate students in and outside the department.

The interpretation of the information gathered through this fieldwork is informed by a second type of research, which might be glossed as "archival." More accurately, this second type of research involved gathering and analyzing information from a variety of scholarly and nonscholarly sources, often areas that seemingly had little to do with epigenetics. Before, during, and after fieldwork, I compiled news articles, scientific articles, policy and institutional documents, and more informal reflections on infertility and the environment, reproduction and hormones, food and exposures. I also documented conversations with Nanjing locals and observations of everyday life outside the laboratory setting that overlapped with these issues. This information has influenced my interpretation of epigenetic research practices, and some has even been directly incorporated in order to help readers make sense of the contingency of epigenetic research. By bringing together this wide variety of sources, my book explores the ways that doing epigenetic research in Nanjing, during a moment of growing but still largely aspirational attention to environmental pollution, shapes how epigenetic knowledge is produced.

These varied sources and observations were helpful in thinking through why the DeTox Lab's research materialized certain environments and not others. Moving through a variety of research sites—from media to nongovernmental organization (NGO) reports, from hospitals to laboratories, from transnational scientific literature to everyday scientific practice—also gets me away from an idea that the boundaries of the laboratory are clear or that the laboratory is a site of replication (Knorr-Cetina 1999) rather than an environment that itself comes into being in relation to various other environments. Even when this book focuses on what has been called "science in action" (Latour 1987), it also points to the material and existential circumstances—political and industrial histories, institutional infrastructure, gender and ethnic stereotypes and expectations—that shape epigenetic research practices. These circumstances are not the contexts of science; they are a part of scientific practice and inform the scales at which environments are brought into being.

Organization of the Book

This book is organized around five environments that were prominent in the DeTox Lab's research and in conversations and media representations that occurred in and outside the lab. Each environment circumscribes a set of material-existential circumstances of interest to the toxicologists I studied and addresses the threat of living with the embodied consequences of toxic exposures for present and future generations. Each chapter is meant to encourage readers to think differently about what it means to conceptualize and materialize "the environment" through epigenetic research. What does the environment mean to those who study its connection to reproductive health, especially Chinese men's reproductive health? In a moment of increased scientific and activist attention to the environment, who becomes responsible for interconnected, intergenerational health? Each of the book's five chapters offers different answers to these questions.

But—like the environments that the DeTox Lab studies—the chapters also overlap. The national, hormonal, dietary, maternal, and laboratory environments are made to stand apart even though they are partially connected. Materializations of various environments are a means through which toxicologists at the DeTox Lab both oversimplify causality and come to understand relationships between economic, social, industrial, and dietary transformations in men's reproductive and developmental health. The book begins with the DeTox Lab's earliest research and questions of sperm decline, moves on to the human and animal studies that occurred during my fieldwork, and ends with a discussion of their later birth-cohort studies. Despite this linear rendering of time, much of this research, especially in the later chapters, occurred simultaneously. As explored further in the coda, this organization of the text is meant to allow for reflection on how the goals of the both the DeTox Lab's research and the transnational study of environmental health have changed over time as environments increasingly proliferate in and through intergenerational environmental health research.

Chapter 1, "The National Environment," begins in 2005, when the global sperm crisis "washes up on China's shores" ("Sperm Crisis" 2005). Originally articulated by a research group in Denmark in 1992 (Carlsen et al. 1992), the term *global sperm crisis* signaled a decline of sperm counts and quality over the past fifty years that was thought to be linked to environmental change the world over. Debate about such claims resulted in another hypothesis: that sperm decline was not global but instead was a matter of geographic variation. The DeTox Lab researchers extended both hypotheses when they began conducting "toxicogenomic" research on the potential intergenerational effects of damaged sperm DNA in the occupational environment. Though to some extent participating in a kind of DNA fetishization (Franklin 1988; Haraway 1997a), their research highlighted DNA's vulnerability to geographically specific environmental factors, pointing to the workplace as a site through which the embodied effects of China's unique role in the world economy could be understood. Moreover, this initial study's findings reverberated with growing national concern about "population quality" (*renkou suzhi*) amid restrictive birth-planning policies. Subsequent research on men's infertility among the "general population" showed that everyday exposure levels among infertile men in Nanjing were many times higher than those reported from other national settings. This finding created a transnationally comparative lens through which a toxic national environment came into view. Finding a toxic "national environment" as partially responsible for semen decline, the DeTox Lab articulated the embodied and intergenerational consequences of China's industrial pollution and lax environmental regulations through the framework of genotoxicity.

Chapter 2, "The Hormonal Environment," discusses endocrine-disrupting chemicals (EDCs), which are objects of study in much epigenetic toxicology, through an analysis of a 2010 Greenpeace China report titled *Swimming in Poison*. The report was referenced by toxicologists from the DeTox Lab and was modeled on previously conducted toxicological experiments, but it took on the challenge of making EDC toxicity a matter of public concern. Quite unsurprisingly for the toxicology community, the report found that fish from collection points along the Yangtze River (*Chang Jiang*) showed elevated levels of harmful "environmental hormones" (*huanjing jisu*, also often referred to as EDCs). Scholars have critiqued EDC science and activism for its heteronormative pathologizing of reproductive and developmental harm, drawing attention to the "sex-panic" that emerged around EDCs' "gender-bending" effects. This chapter shows that such sex panic is not necessary for activist success, nor is it a universal obsession in responses to EDCs. Unlike in Europe and North America, media reactions to the report in China did not focus on sex transgression. Instead, reactions focused on food safety, industrial capitalism, and the ecological scope of pollution. Based on this analysis, I argue that the analytic potential of the hormonal environment, and of toxicology more generally, might be better mobilized through cultivating attention to underlying social, political, and economic causes rather than through panic over harmful effects.

Epigenetic research often focuses on the way diet influences the health of future generations, drawing attention to the intergenerational impacts of "food as exposure" (Landecker 2011). The effects of phytoestrogenic plants, or plants that contain EDCs, are of particular concern to reproductive and developmental toxicologists. In chapter 3, "The Dietary Environment," I think through studies of soy consumption and its disputed influences on men's reproductive health, particularly the health of sperm. I show how a US-based study that found negative influences of soy-consumption habits was received in China and motivated the DeTox Lab to launch its own investigation of soy-sperm relations. This comparative study of soy brought a "Chinese body" into being, said to be distinguishable through dietary habits, metabolic capacities, and genetic polymorphisms. In its research on the dietary environment, the DeTox Lab attempted to counter research assumptions about the negative effects of phytoestrogens on men. Through challenging such research, the DeTox Lab also challenges stereotypes of femininity and masculinity historically aligned with East and West. But they also end up reasserting a familiarly deterministic idea of race as genetically definable, even if interindividually variable, resulting in the racialization of the metabolism. Through its implicitly comparative dimensions, the dietary environment becomes as materially and semiotically fixed as "Chinese men."

Chapter 4, "The Maternal Environment," turns from the laboratory to a partnering hospital to consider the argument commonly made by social scientists that epigenetic research exaggerates maternal blame for inherited conditions. Drawing on fieldwork in a neonatal unit that treats congenital disorders and participates in toxicological research, I show that epigenetic studies of infertility and congenital disorders conducted by the DeTox Lab encourage physicians and patients to deindividualize ideas of maternal responsibility. Toxicologists bring into being a maternal environment that both reasserts maternal responsibility for fetal health and places responsibility on intergenerational human and nonhuman kin, thereby reconfiguring preexisting models of relational personhood to reassert a sense of intergenerational connectivity and collective responsibility. I argue that the understanding of personhood underlying critical social scientific critiques of the maternal environment often relies on a Eurocentric model of personhood and misses the potential of epigenetic research to interpret the person from a relational perspective.

Environmental epigenetic research has been praised by many for its complex approach to genes and biology. But such research has also been criticized for its tendency to reduce complex activities into oversimplified characterizations of environments. In chapter 5, "The Laboratory Environment," I explore the laboratory space in which the environment is both reduced and proliferated. I show how DeTox Lab members demarcated, isolated, and measured the influence of environmental factors in and on animal models, while simultaneously observing the multiple exposures and complex contexts in which all animals live, eat, breathe, interact, and reproduce. In such work the laboratory environment itself becomes one of many environmental factors thought to have epigenetic influences. Through ethnographic depictions of experiments conducted by DeTox Lab members, I show how situating the laboratory in its social, cultural, and environmental settings is not the exclusive analytical purview of science and technology studies. It is also the increasingly necessary work of those who study how environmental factors potentially influence the results of intergenerational animal experiments in the laboratory environment.

In the coda I reflect on the DeTox Lab's birth-cohort research that occurred after I left China and discuss the rise of intergenerational research in environmental health more broadly. I suggest that as a contextualizing force that places individual health conditions within broader social, political, economic, and chemical conditions, epigenetic toxicology and birth-cohort research more generally have the potential to rewrite the boundaries of the reproductive body, deindividualizing reproductive and environmental health responsibility in China and beyond. But they also have the potential to reify reproductive norms and biologically deterministic ideas of race and kin. The epilogue is a short reflection on my own struggles against the individualized burden of reproductive responsibility through a discussion of breastfeeding jaundice and the paradox of plastics.

Reproducing a Toxic China

As toxicologists bring environments into being through epigenetic research, emphasizing the elevated levels of exposure faced in the environments that surround them, a narrative of a toxic China emerges. The story is in many ways familiar. Over the last three decades, after the death of Chinese Communist Party leader Mao Zedong and the embrace of reform and opening policies that integrated China into a global economy, the nation rapidly industrialized. Now the world's largest economy, China is dealing with an economic boom that has come at great costs. Unbridled industrialization and a lack of environmental regulation often characterized as a policy of "pollute first, clean up later" have led to rampant air, water, and soil pollution. In more recent years this pollution has been coupled with the rise of consumption among a growing middle class that is increasingly purchasing products such as automobiles, meat, and other commodities once viewed as luxuries. The growing pollution of China's environment has become a serious burden for its residents, who suffer from environmental health problems, and for the national government, which now dedicates a significant percent of its annual budget to environmental remediation.

The sense of toxic ubiquity that this narrative evokes leaves environmental health scientists and activists in China, as well as anthropologists who study them, in a position of foregrounding Chinese toxicity in ways that are both important and potentially problematic. On the one hand, the scale of China's toxicity is undeniable. Beneath the paradoxical narrative of China's toxic transformation, people's lives are being upended and ended by a rising number of environmental health concerns. There are now more than four hundred "cancer villages" in the People's Republic of China (PRC) and twenty-five in Jiangsu Province alone (Cheng and Nathanail 2019). A growing number of health concerns are now considered issues of environment health, from cancer to cardiovascular disease, Alzheimer's to low birth weight (Holdaway 2013). These are urgent concerns for Chinese people and increasingly for the PRC government. On the other hand, scholarly and media representations of a toxic China have their own effects.

Anthropologist Ralph Litzinger and interdisciplinary scholar Fan Yang characterize foreign media's obsession with China's ecological disasters as a discourse of "Yellow Eco-Peril." Such discourse depicts China as "a polluting and polluted Other" (Litzinger and Yang 2019, 211). Such depictions depend upon past colonial characterizations of China as the "Sick Man of Asia" (Rogaski 2019). Similarly, Mel Chen shows how an obsession with the toxicity of products made in China problematically racializes toxicity (M. Chen 2012). Such toxic imaginaries are created through a comparative lens, but they also, as Chen as well as Litzinger and Yang note, often decontextualize China from its global economic surroundings, leaving the responsibility for Euro-American consumption and capitalism off the table.

It is then essential that toxicological claims, and the anthropologist's portrayal of them, do not further an idea of toxicity "made in China." The urgency of avoiding such a portrayal has only heightened amid the global spread of the novel coronavirus, COVID-19. (As I write this, the pandemic unfolds, as does a wave of hate crimes against Asian and Asian-American people in the United States and other locations because of a debate over virus origin playing out at the transnational level.) In this book I try to prevent essentializations of Chinese toxicity—not by avoiding the rhetoric and realities of pollution and its effects but by trying to understand how environmental health scientists and activists in China think about and materialize scientific evidence about such issues. I take seriously the findings of those I studied at the DeTox Lab, who demonstrate that people living in the Yangtze River Delta are exposed to more toxins at higher levels than are people in many Euro-American settings. But I also reflect on the presentation of toxic environments and conditions by toxicologists and other environmental actors. By studying the scientific practices in which findings of toxic exceptionalism are materialized and circulated, I hope to approach toxicity as a means of understanding both the discourse and the chemistry of environments.

Accordingly, the title of this book, *Infertile Environments*, is not meant as a description or prediction of a future looming ahead for China. Instead, it is meant to capture an increasingly common anxiety about the connections between environments and the in/fertility of present and future generations in China and beyond, an anxiety that is rooted in both rhetoric and reality. In subsequent chapters, each focused on a particular environment, I show how epi/genetic research provided an avenue for the DeTox Lab to further stress the crucial role of "the environment" in reproductive health. At a moment when concern about toxicity in China was growing but the government had yet to implement consequential monitoring, regulation, or limitation of industrial pollution, the DeTox Lab used bionormative and heteronormative epigenetic research approaches to explore the inheritable dimensions of economic policies that drive national and regional industrial pollution. Their research provided evidence of the comparatively high levels of toxic exposure endured every day by people living in China. It also brought to the fore an understanding of environments as interior and exterior settings that influence health within and across generations. The title is also, then, meant to speak to my sense that epigenetic research is a flawed but persuasive means of exploring how environments outside the body influence men's reproductive, developmental, and intergenerational health.

ONE

The National Environment

In the early 1990s a team of Danish andrologists presented at an international workshop sponsored by the World Health Organization about environmental influences on reproductive health. At this workshop, and in a now widely cited paper, Elisabeth Carlsen and colleagues argued that sperm counts had dropped by 40 percent worldwide over the past fifty years (from 1938 to 1990). Based on semen-analysis results from more than twenty countries across five continents, they further argued that these falling sperm counts were likely not caused by genetic factors. Instead, they suggested, sperm decline was likely associated with prenatal exposures to estrogens, estrogen-like compounds, or "other environmental or endogenous factors" (Carlsen et al. 1992, 612). In the discussion of their findings, Carlsen and colleagues focus on the urgent need to conduct further research on the possibility that changing material and chemical conditions were responsible for falling sperm rates that could be documented through longitudinal research.

Research published prior to the 1992 study, particularly within the fields of industrial hygiene and occupational health, had shown specific connections between workplace exposures and sperm decline (Bohme 2015; Daniels 2006). But the scale of claims made in the Carlsen publication went beyond the occupational environment to suggest that exposure to estrogen-like compounds and environmental pollutants, which had increased in the last five decades, was causing a more global sperm decline. According to their research, semen decline was not confined to those working in dangerous conditions; all men's sperm were at risk. Moreover, sperm decline was not isolated to a particular country or region but instead was a worldwide phenomenon. By characterizing sperm decline as global, in more than one sense of the term (Franklin 2005), Carlsen and colleagues implied that something about this moment in history had changed sperm for the worse. That something—they argued—was not a matter of genetics but a matter of environmental influences.[1]

Since these initial claims about global sperm decline, men's reproductive health specialists from various locations have reported falling sperm counts. These reports have been captured and sensationalized by media headlines and articles that foreshadow the human species' demise via a global sperm crisis (see Ghosh 2017). The now commonly perceived sperm crisis functions through a narrative similar to other crises of the twentieth and twenty-first centuries (Roitman 2013). As described by anthropologist Joseph Masco (2017), the crisis narrative is about engaging the future through negative effects. In an infertile future, endangerment and even extinction loom as threatening possibilities when reproduce-ability is diminished, giving way to a future in which all reproduction is assisted (Lamoreaux and Wahlberg 2022). Such a future is foretold in other species as seeds of various plants and the sperm of nonhuman animals are gathered and stored in banks around the world awaiting their revival via reproductive technologies. Even human sperm is now being stored in cryopreservation chambers as insurance for an infertile future yet to come (Wahlberg 2018a). Through the lens of a sperm crisis, preserving and protecting men's reproductive health becomes a matter of preserving mankind, gendered emphasis intended (Daniels 2006).

In China the narrative of a global sperm crisis has taken a slightly different tone. Here, the sperm crisis (*jingzi weiji*) has been interpreted as an outside force sweeping the globe and now "washing up on China's shores" ("Sperm Crisis" 2005). But sperm decline has also been perceived as part of a more localized "sperm bank emergency" (*jingzi ku gaoji*). National sperm shortages have repeatedly been declared by sperm banks unable to meet growing demand for donor sperm to be used alongside assisted-reproductive technologies in various regions. Like the global sperm crisis, the sperm bank emergency is fundamentally about the potential inability to reproduce. But it also characterizes semen decline through nationalist logics and within geopolitical boundaries. Unlike the sperm crisis, which is perceived as a global threat, the sperm bank emergency is presented as a national problem—both of reproducing persons and of reproducing the Chinese nation itself.

Genotoxic Geographic Variation

Following the 1992 Carlsen study, andrologists in countries around the world debated the claim that environmental factors were causing global sperm decline. Many found the research convincing and prepared to conduct similar studies, but others criticized the methodology of Carlsen and colleagues' original research (Daniels 2006). More than thirty longitudinal studies of sperm decline were published in the fifteen years that followed, many of which confirmed falling sperm counts in their respective settings. Such settings were most often portrayed at the level of the nation; "rates of national sperm decline" quickly became one of the most common frames of analysis. As Cynthia Daniels explains, "A nation's sperm count was a measure of its national virility . . . Finnish men could 'stand tall' while American men 'faced extinction'" (2006, 59).

Amid a wave of attention to regional sperm bank emergencies in China, the then nascent DeTox Lab took up the question of geographic variation in sperm counts as well. However, its research was motivated less by concerns about the national virility of Chinese men and more by the possibility that socioeconomic and environmental change was negatively influencing Chinese men's ability to reproduce healthy offspring. The DeTox Lab began in the early 2000s, when a recent PhD graduate and new assistant professor, Zhang Zhiyuan, began to design research that would investigate the effect of China's burgeoning industrial chemical production on men's reproductive health. At this time, China was the world's largest user, consumer, and producer of pesticides (Hamburger 2002). The Yangtze River Delta region, where the Nanjing Institute of Medicine and Science and the DeTox Lab were based, was home to many pesticide and other chemical factories, along with many workers potentially facing regular, low-dose exposures to various toxins. In his new role as lead researcher, Zhang pursued research that would not only articulate China's position in the supposedly global sperm crisis. His research would also ask important questions for his region and his country about the indirect and intergenerational effects of toxic exposure in the occupational setting.

In particular, Zhang was interested in the toxicogenomic effects of exposures, as were many toxicologists at the time (Fortun 2011). According to environmental historian Scott Frickel, US-based toxicology embraced "genotoxicity" in the late twentieth century as a means of describing and documenting the genetic damage associated with synthetic chemical exposures. The narrative of genotoxicity stated that "damage to an individual's sex cells initiated by environmental mutagens could, if passed from parent to offspring, remain within the population for generations and ultimately compromise the long-term integrity of the human gene pool" (2004, 1–2). In other words, the threat of genotoxicity was mobilized as a vehicle for furthering regulatory change under the guise of improving multigenerational population health and longevity. Such framing of environmental damage as genetic, therefore potentially inheritable, resulted in increased scientific attention to the potential intergenerational consequences of synthetic chemical exposures. Frickel suggested that seeing genetic disease in the environment, and the environment in genetic disease, was a turning point in thinking about the causal factors of genetically related health. In his words, "Genetic disease and environmentally induced mutations were posed as mirror images of one another" (2004, 106). However, such mirroring of genetic disease and environmentally induced mutations was not only compelling but also surprising at this particular moment in Euro-America because genes and the environment had come apart (Strathern 2005).

The idea that environmental mutagens threaten genetic health would have been less surprising in China, where the Chinese government's late twentieth-century emphasis on cultivating "superior birth and child rearing" (*yousheng youyu*) focused on how genes *and* environments shape humans in formation. Through policies related to such cultivation, genetic makeup and constitution at birth were viewed as starting places for ensuring the health and quality of the nation and its population. This led to a kind of positive eugenics that stressed best practices for cultivating children as both "well-bred and well-reared" (Jiang 2015).[2] In the 1990s a related campaign to improve "population quality" (*renkou suzhi*) was underway and accompanied by the introduction of a family-planning policy that limited most urban families to one child.[3] These policies were justified by the Chinese Communist Party as the correct path toward the country's modernization, transforming the "backward" masses into a population lower in number and higher in quality (Anagnost 2004; Greenhalgh and Winckler 2005). This confluence of policies exacerbated pressures on parents, particularly women, to have and rear "high-quality" children. In anthropologist Susan Greenhalgh's words, "With but one, that one had to be perfect" (2010, 58).

Reproductive health specialists in China had previously framed applications for state-sponsored scientific funding through the goal of improving population quality (Jiang 2015; Wahlberg 2016).[4] Amid such reproductive governance (Morgan and Roberts 2012), the DeTox Lab's research similarly worked through and fostered concern about the potential generational effects of toxic exposures on the national population. Highlighting the vulnerability of DNA to polluted environments, the DeTox Lab communicated the national importance of their research. Its studies suggested that the effects of genotoxic chemicals on sperm went beyond a single individual's fertility to inform the developmental and reproductive health of offspring in future generations. This focus on the intergenerational threat of toxic exposure increased the lab's national and international reputation, allowing its research on male infertility and other reproductive and developmental health issues to continue through government funding, even while it often implicitly critiqued a national lack of environmental regulation.

The Occupational Environment

The DeTox Lab's research on sperm decline and male infertility, conducted during a wave of attention to the possible environmental causes of a sperm crisis, was both a response to global sperm crisis discourse and distinct from these concerns. Zhang and a small number of graduate students traveled from the capital of Jiangsu Province, Nanjing, to the eastern city of Changzhou. Like many cities in the region, Changzhou has an adjacent industrial zone that sits along the river. At a factory in this large zone, the group of researchers led by Zhang conducted research on the effects of regular, low-dose exposure to pesticides in the occupational environment.

Inside the Changzhou pesticide factory, DeTox Lab members worked through methodologies established in the toxicological tradition of studying the occupational environment (Sellers 1997). The team monitored the air quality over three days, measuring the concentration of specific pesticides in the factory's atmosphere. They conducted exposure assessments on a few individuals per day, using dermal measurements as well as active personal samplers—small devices that measure contaminants in air. They also collected semen samples from approximately fifty plant workers—some from within the factory and some from an office nearby. Finally, they conducted general health evaluations of all participants and took their medical and occupational histories. These records and biomaterials were brought back to Nanjing for analysis, standing as physical and informational representations of the effects of "the occupational environment," a demarcated space in which bodies and pesticides come together through regular, low-dose exposures.

After collecting samples from the Changzhou factory and other control settings, members of the DeTox Lab analyzed the collected semen and blood. They performed routine semen analyses, in accordance with the guidelines published by the World Health Organization (WHO), and found that chronic, low-dose exposure to multiple pesticides was associated with diminished semen quality, inducing changes in sperm counts, motility, and mobility. But it was further DNA analyses of the collected sperm that led to what they considered an even more important finding. The team used imaging techniques to obtain information on the degree of chromosomal aberrations, DNA fragmentation, and DNA strand breaks in the collected sperm. In a series of investigations the team found that sperm's "DNA integrity" was lower in the occupationally exposed population than in the sperm from men working in surrounding office settings or men in an external control group, collected off-site. Most importantly, they found that even when traditional semen analysis results indicated a "normal" specimen, the sperm from men working on the factory floor had lower DNA integrity than the other research participants.

Results from the DeTox Lab's research were eventually presented in a series of publications which argued that pesticides produced in the factories they studied had "genotoxic" effects on human sperm. Publications emphasized the potentially harmful effects of pesticides as well as the importance of considering DNA damage in sperm evaluations. DNA damage did not necessarily lead to male infertility. On the one hand, this made such damage less worrisome. Pesticide factory workers were being exposed to levels of organophosphate compounds, pyrethroids, and carbamates at higher levels than recommended by the US Occupational Safety and Health Administration, as the DeTox Lab's research notes. Still, infertility had not been a common experience among men in this occupational setting.

On the other hand, a lack of infertility did not mean that the sperm of pesticide factory workers was not negatively influencing the reproductive and developmental health of future generations. Instead, it meant that although sperm could still fertilize, it might also transmit "damaged" genes. The effects of genotoxic exposures on sperm potentially went beyond a single individual's fertility to inform the developmental health of offspring in future generations. Based on this analysis, the DeTox Lab argued that the evaluation of sperm DNA was as important as routine semen analysis. They stressed that certain semen-analysis techniques had become routine and that, following the guidelines of the WHO, these techniques were now obligatory for international scientific publication and research standardization. Yet these standardized analyses did not provide a full picture of the potential effects of the occupational environment on sperm. The DeTox Lab's research challenged the taken-for-granted idea that routine semen analysis should not include considerations of DNA, and the more implicit idea that sperm DNA is impermeable to the environment, occupational or otherwise.

Denaturalizing Sperm Selection

There was another layer to the DeTox Lab's focus on DNA damage. At the same time that the lab's sperm research was being conducted and industrial pollution was increasing, the availability of assisted reproductive technologies (ARTs) was expanding in China. In the early 1990s anthropologist Lisa Handwerker conducted research on the growing use of ARTs in Beijing. The first in vitro fertilization (IVF) baby in the country was born in 1988, and that number expanded to fifty within the first five years of availability. Handwerker argues that the growth and overwhelming demand for biomedical fertility services was a result of several factors, including "a long-standing cultural imperative for women to reproduce (ideally, sons)" (2002, 301). This imperative was exacerbated by restrictive birth-planning policies, which were a painful reminder for infertile women, laying bare "the assumption that all women of reproductive age should be fertile" and contribute to the continuation of patrilineal kin (2002, 302). In addition to such gendered cultural expectations, Handwerker points out that reproductive technologies were highly profitable in a newly commercialized medical economy. Many fertility clinics were making money by offering expensive "Western" technologies, as "test-tube babies" (*shiguanying'er*) were touted as miracles that changed the lives of previously infertile women.[5]

Even in a landscape of medical profiteering, lack of accessibility to biomedical fertility treatment was a major issue in China; fewer clinics existed to serve a large population, and IVF, which was not covered by medical insurance, was extremely expensive for most people (Qiao and Feng 2014). Government regulations also began to more strictly regulate the arena, determining who could receive fertility treatments and of which types. In the early 2000s the Ministry of Health instituted new policies on reproductive technologies, including the Regulation on Human Assisted Reproductive Technology (*Renlei Fuzhu Shengzhi Jishu Guifan*) and Ethical Principles for Human Assisted Reproductive Technology and Human Sperm Banks (*Renlei Fuzhu Shengzhi Jishu he Renlei Jingziku Lunli Yuanze*). These regulations had at least two aims. First, they were put in place to establish control of the growing number of biomedical fertility clinics. (By 1999, there were approximately two hundred such clinics across China, many of them unregulated [Wahlberg 2018b].) These new regulations would approve and monitor fertility clinics, most of which were affiliated with state-run hospitals (Qiao and Feng 2014). Second, many requirements were established to ensure that reproductive technologies would contribute to national birth-planning policies and population goals.

In order to adhere to national birth-planning policies, only certain couples were allowed to undergo treatment through reproductive technologies. At the time of my research, in order to undergo infertility treatment through reproductive technologies, patients had to present marriage and birth-permission certificates. The Mother and Infant Health Care Law (*Muying Baojian Fa*) prohibited those with a long list of conditions deemed to be undesirable and/or inheritable from using reproductive technologies, including sexually transmitted disease, serious genetic diseases, drug addiction, and exposure to teratogenic radiation (Qiao and Feng 2014). Furthermore, certain treatment options were banned, including pre-implantation genetic diagnosis (PGD) for gender selection and surrogacy. Egg and sperm donation was allowed under conditions of anonymity and strict screening guidelines, but compensation for donated gametes and directed gamete donation was not (Qiao and Feng 2014). Physicians were able to transfer only two or three embryos, depending on the age of the woman undergoing treatment, to limit multiple births.[6]

Another related difference in the use of reproductive technologies was the way they interacted with general anxiety about congenital disorders, partially brought about by maternal pressure to birth and rear "high quality" children (J. Li 2020). This is where the DeTox Lab's critique of reproductive technologies began. One specific technique, intracytoplasmic sperm injection (ICSI), had been studied for a potential association with increased risk of congenital disorders or "birth defects." ICSI involves the selection and injection of a single sperm cell into the cytoplasm of an egg with a small needle. This technique does not and currently cannot test the specific sperm cell it injects for DNA integrity because the tested sperm is destroyed in the process. Therefore, some worry that "paternal subfertility with a genetic background" may result in increased risk of congenital malformations, particularly hypospadias, when ICSI is used (Ericson and Källén 2001).[7] In a context of growing ICSI use, public health advocates began using accounts of genotoxicity to question infertility-treatment methods. Such concern is commonly expressed by men's reproductive-health experts as a matter of "bypassing natural sperm selection." In other words, instead of sperm competing for first entry into the egg, humans are selecting sperm for the qualities that they have deemed most likely to successfully fertilize. Studies of sperm DNA damage and its potential developmental and intergenerational effects highlighted the risk of such denaturalized sperm selection, where nature is understood as outside the human.[8]

DeTox Lab scientists interpreted the threat of ICSI a bit differently. They did not advocate removing the human element from a "natural" sperm selection practice but instead improving upon an already denaturalized sperm selection process. Because sperm and the DNA it contained were perceived as influenced by the environment, the way forward was not eliminating ICSI, but studying the environmental factors that contributed to potential genotoxicity. Such thinking was key to the lab's reframing of the causal factors of male infertility. The lab saw sperm as always already influenced by what stands outside it, including human prioritization of industrial development and the nonhuman substances involved in such pursuits. The DeTox Lab's framing questioned the very idea that sperm was simply a "natural" substance in the first place. Cutting through the nature/culture, biology/technology debates long emphasized in Euro-American biology and genetics, the DeTox Lab problematized the idea of natural selection via sperm through a shift to the potential chemical roots of reproductive and developmental toxicity.

Biomarkers of the Nation

The DeTox Lab's early studies of the occupational environment raised questions about the harm of pesticides even in the absence of infertility. Its next set of studies shifted to infertility directly and broadened the lab's scientists' conceptualization of the environment to a national scale. Inspired to investigate both the hypothesis of endocrine-disrupting chemical (EDC)–related sperm decline and its geographic variation, in 2004 the DeTox Lab began researching the effects of toxic exposures more common in low-resource countries such as China. They recruited research subjects from a pool of infertility patients at a local hospital and worked with physicians and staff to organize the collection of various samples from hundreds of men, including semen, blood, and urine. These samples were anonymized and transported to the Nanjing Institute of Medicine and Science (NIMS) Toxicology Department, where they were stored in large freezer chests, awaiting analysis. The team also collected biographic, dietary, lifestyle, and occupational information from participants, and conducted basic physical exams.

The DeTox Lab was particularly interested in investigating the potential connection between infertility patients' exposure to EDCs and reductions in sperm quality and quantity. Studies of EDCs demand a significant shift in both toxicological theory and method because historically the discipline has focused on exposure thresholds to individual chemicals (Murphy 2006, 2013; Langston 2011). However, EDCs have been shown to have effects with chronic exposure at low doses. They also can have different effects on different bodies that are more or less susceptible, and when combined with other chemicals. In order to assess a large group of men's exposure levels to various EDCs, and test for their potential relationship to infertility, the DeTox Lab's new research project estimated toxicity through internal biological markers of exposure, more commonly referred to simply as biomarkers.

Biomarkers are an increasingly common tool used within the natural (and sometimes social) sciences. Although biomarkers are understood as changes in the biological systems of animals, they are interestingly defined not by what they are but by how they are used. According to a toxicologist writing in the late 1980s, "The term 'biomarker' refers to the use made of a piece of information, rather than to a specific type of information" (Henderson et al. 1989, 65). Biomarkers are thus symbolic in the sense that they are not useful until they come to stand for something other than themselves. In toxicology, biomarkers are used to measure both exposures and effects; they are biological traces left by exposures or the effects of exposures. Using biomarkers allows researchers such as those at the DeTox Lab to study diffuse, low-dose, and ambient exposure to EDCs.

One of the primary biomarkers used by the DeTox Lab was urinary metabolites. Today, metabolites are viewed as more than the physical aftermath of exposure, internally processed then excreted by the body. Like the metabolism itself (Landecker 2013), metabolites are now understood as information. They are "internal" measures used to estimate levels of exposure from "external" sources. Urinary metabolites are used to estimate levels of exposure to toxins that would otherwise be difficult to assess because of diffuse or multiple exposures. In the lab's research, urine samples from male infertility patients were thawed and analyzed in a mass spectrometer for metabolites of particular EDCs previously shown to have adverse effects on sperm. These pesticides were thought to have been taken in by research subjects from multiple sources that would otherwise be difficult to trace, such as food and drink residues or dust within homes. By measuring specific metabolite levels in urine, the lab estimated exposure levels to specific pesticides and other toxic chemicals, then analyzed the relationship of these levels to semen parameters. Through this metabolic technique the DeTox Lab researchers contributed multiple studies to global debates about the cause of sperm decline. Importantly, they found that high urinary metabolite levels of certain pesticides that had been classified as EDCs slowed sperm progression and vigor. Based on these findings, EDC exposure was associated with sperm decline among what they referred to as the "general population."

But perhaps even more important to the DeTox Lab's research was another discovery made along the way. When writing up their research, the DeTox Lab scientists began to compare the levels of EDC-linked urinary metabolite concentrations found in their research subjects to those in other national contexts. They found that some pesticide metabolite levels were four times as high in their research subjects as they were in the research subjects of studies based in the United States and up to eight times higher than in European research subjects. Meanwhile, research subjects in other regions of China, including Beijing, showed similar levels to their own. Through such comparisons, it was becoming clear that China's "general population" faced greater exposure levels and higher likelihood of sperm abnormalities than the general populations of other countries. Through this comparative lens, the level of exposure to EDCs—even without a consideration of effects—became a matter of national public health, the DeTox Lab argued.

To be clear, the DeTox Lab's findings were not based on representative samples from throughout China. Their studies did not measure levels of EDCs or sperm quantities or qualities throughout the country or even throughout Jiangsu Province, where Nanjing is located. But because of exceptionally high levels of metabolites among infertility patients in Nanjing as compared to infertility patients in other countries, the environment negatively influencing sperm was conceptualized through a national frame as a study of the general population in China. Here, externalization of information via biomarkers geographically differentiates sperm decline. Such geographic differentiation is not a matter of innate biological difference but of chemical exposure. The DeTox research thus brought to the fore a crucial issue bypassed by other sperm scholars, even those concerned about environmental factors: the uneven global distribution of EDCs and human exposures.

Local, Exposed, and Situated Biologies

Much of the DeTox Lab's research, and gene-environment interaction or epigenetic studies more generally, has similarities to the popular anthropological concept of *local biologies* (Lock 1993). Lock's original discussion of this term came about after fieldwork on women's aging in Japan, during which qualitative and quantitative data described women's aging experiences as remarkably less burdensome than those typically described in Europe and North America. Lock suggested that such difference was not just a matter of the cultural interpretation of a single, universal biological phenomenon: menopause. She theorized instead that biology itself was shaped by its contexts (Lock 1993).

If one thinks of environments as contexts, one might certainly interpret the work of the DeTox Lab as about local biologies. Its research is centered around the high levels of toxic exposure that are faced by those who live and work in the region where the lab is located. Building on Lock's concept, anthropologist Ayo Wahlberg has theorized such high levels of localized exposure in China through the idea of "exposed biologies" (2018a).[9] Exposed biologies focus on how "industrially manufactured chemicals and modern forms of living are increasingly held culpable for a range of pathologies, from cancers to metabolic diseases, respiratory troubles, cardiovascular conditions, to disorders of sex development and infertility" (Wahlberg 2018a, 309). Moreover, the concept of exposed biologies foregrounds a perceived need to protect the body from what stands outside it, a practice that Wahlberg sees in the selling of fertility insurance in China, which involves the cryopreservation of human semen and the storage of this fluid in airtight chambers sealed from the outside world.

For Wahlberg, such practices show that exposed biologies are a rhetorical vehicle through which the routinization of technological solutions to complex social and environmental problems occurs. Biology itself is interpreted here by cryopreservationists, if not Wahlberg, as newly exposed and in need of protection at a specific historical moment in China. The DeTox Lab's researchers' rendering of the biological is a bit different. For them, it is not so much that biology is newly exposed in what Wahlberg calls "China's Anthropocene," defined as "a historical moment when anthropogenic effect was being detected, not so much in atmospheric carbon dioxide levels, glacial ice cores or sea levels, but rather in levels of carcinogens, mutagens, obesogens, teratogens and endocrine disruptors found in the human body" (2018a, 310). For the lab, biology has always been a part of the world that stands outside it. The body is understood as an ongoing practice more than a finished object, fundamentally shifting and intertwined with what exceeds the skin (Farquhar 1991). Biology in China today is thus just as permeable as it ever was, and the environment is just as imbued with human activity as it always has been. It is the increased pollution of the environment that is problematic, not environmental exposure itself. In other words, instead of biology or the body being newly exposed, it remains just as porous. It is, then, environments that are newly polluted.

This distinction is important because it places the responsibility for and solution to threatened reproductive health at a different scale than the solution proposed through individualized fertility insurance. If biology is exposed, biology itself needs to be protected. Man stands apart from the environment and indeed seeks isolation from it. But if environments are polluted, and states of pollution correlate between interiors and exteriors, then the solution to semen decline, for instance, is remedying pollution, both inside and outside the body. In the DeTox Lab's rendering of environmental exposures, biologies are not mere effects of exterior environmental contexts; they are mutual instantiations of circumstances. Mei Zhan calls such correlation "an analytic of oneness" and suggests that the Chinese idea of *tianrenheyi* (insufficiently translated as "heaven and human are one") opens up "the possibility for recuperating and envisioning a worldly future of healing and being oriented to the undividedness of the human and the world" (2011, 108). According to Zhan, an analytic of oneness demands that we rethink what it means to be human. Perhaps it also demands we rethink what we mean by environment and how local contexts are not simply around us but also within (Strathern 2013).

Lock's idea of local biologies argues that both bodies *and* ways of knowing bodies coexist with their local contexts. In later works she shifted, in collaboration with Jörg Niewöhner (2018), to the term *situated biology*. This shift was an attempt to update local biologies to the movement of today's world and a nod to Donna Haraway's (1991) concept of situated knowledge. Unlike local biology, situated biology draws attention to the shifting influences of power in shaping both contexts and knowledge about those contexts. In the case of the DeTox Lab, this means that biology (measured via sperm) expresses its polluted environments and potentially reproduces environmental effects in future generations. But it also means that knowledge about sperm and environments is itself situated within the shifting political, socioeconomic, and historical contexts in which it is produced.

Paradoxes of Ubiquitous Exposures

The DeTox Lab was quite successful in publishing their research both nationally and internationally. Findings from the DeTox Lab's research on the association of high levels of exposure to EDCs and on the decline of sperm quality and quantity among infertility patients were published in multiple journals and circulated in environmental health circles both locally and abroad. Their research also garnered the group enough funding to continue to grow and study the harmful effects of toxic exposures. In retrospective interviews I was told that a large factor in such success had been the fact that nearly every urine sample that members of the DeTox Lab tested for EDC metabolites showed significant traces of exposure. Without such consistent exposure metabolites, research would have potentially taken longer, been more expensive, and viewed as less noteworthy. As it was, with exposure levels to multiple chemicals so consistently statistically significant and comparatively high, the DeTox Lab was now imagining and executing multiple future studies that built upon this research.

Members of the DeTox Lab mentioned other benefits to conducting research in China, many of which have been discussed in comparative studies of bioethical standards (H. Chen 2013; Sleeboom-Faulkner 2013; Thompson 2013; J. Y. Zhang 2012). Recruitment of human research subjects was eased by the straightforward access to study populations and high participation-consent rates, always in the 90 percent range. Research groups often had more graduate students than in the United States and could draw on a larger labor pool to collect samples from partnering institutions. In addition, obtaining human-subjects research-protection reviews was fast and simple compared to ethics approval practices reported by peers in other countries, members of the DeTox Lab told me. Together these factors allowed the lab to offer human data on the effects of toxic exposure to EDCs at a large scale more efficiently than their international counterparts.

Zhang—like many others—had a complicated relationship with the paradox of development so often discussed in conversations about China. Amid his growing personal concern about pollution in Nanjing and its effects on himself and his family, he was thriving professionally through the study of toxicity—quickly gaining publications, national funding, and international recognition. But beyond such personal and professional advance, he saw a country stagnant in its environmental regulation. In his view, his research needed to move forward, making the most of national conditions of exposure, so that the lab he now directed could, as he told me during our first in-person interview in 2011, "research more chemicals, more populations, more exposures."

But for the moment, Zhang and his colleagues at the DeTox Lab saw sperm decline as an entry point to consider how global shifts in chemicals and materials were affecting men in China. Genotoxicity and other forms of gene-environment interaction research were a way for Zhang and other members of his growing research group to think through the consequences of changes at many scales, further situating biology.

TWO

The Hormonal Environment

The first time I met Professor Zhang Zhiyuan, the lead researcher of the DeTox Lab, in person was in January 2011. It had been over a year since we had first communicated by phone during a previous visit to China, and today Zhang was waiting for me at the campus's main gate. Shortly after he greeted me, he directed us across the wide, busy street that ran in front of the institute's south side. From there we walked to an Italian restaurant located in a strip mall that I would later come to know as a popular entertainment area. Inside the restaurant we sat across from each other at a booth, formally introducing ourselves before selecting our individual entrées—an ordering style that was uncommon in China and seemed to be a small gesture of cosmopolitan hospitality. Zhang then set up his laptop at the end of the shared table and began to deliver a PowerPoint presentation about the DeTox Lab's research to just me, an audience of one.

Zhang started the presentation by reviewing how the DeTox Lab's early studies of chemical pesticides in the occupational environment were subsequently scaled up to consider the effects on human sperm of a variety of EDCs found at high levels within China. Zhang then described how, alongside this research, he led a study of how EDCs influence men's hormone levels. This study focused on what toxicologists and other natural scientists often classify as "reproductive hormones," hormones deemed essential to reproductive and developmental processes. Social scientists have observed how such "reproductive hormones" are often gendered, discussed as male or female, even though all humans have and need hormones such as testosterone *and* estrogen for the functioning of many systems (Fausto-Sterling 2000). Evaluating whether or not hormonal levels are within a "normal range" is a method often used by scientists as a measure of "doing sex" properly (Oudshoorn 1994). Hormone levels, in turn, are sometimes used as indicators of reproductive ability. Like many within toxicology, the DeTox Lab subscribed to such hormonal normativity. Their most recent study, Zhang told me, investigated whether "abnormal" levels of so-called reproductive hormones—specifically estradiol, follicle-stimulating hormone, luteinizing hormone, prolactin, and testosterone—were associated with high levels of EDC exposure and semen decline.

While Zhang was describing this research, the food arrived. Two large plates of spaghetti with red sauce were set before us; in the end we had ordered the same dish. Zhang closed his laptop but continued to talk, shifting his tone to a more casual and conversational style. In a mix of Chinese and English he asked me if I had heard of a report called *Swimming in Poison*, which was released by the international environmental NGO Greenpeace (*Lüseheping*) in the late summer of 2010, just a few months before. Zhang outlined the main point of the report: among other alarming data, the report states that fish from collection points along the Yangtze River showed elevated levels of harmful "environmental hormones" (*huanjing jisu*). *Environmental hormone*, I would learn, is a Chinese term that overlaps with the category of EDCs and other English approximations of this chemical class that were once more popular, such as "environmental estrogen" (Krimsky 2000).

The WHO defines an EDC as "an exogenous substance or mixture that alters function(s) of the endocrine system and consequently causes adverse health effects in an intact organism, or its progeny, or (sub)populations" (2013, 4). The number of substances now classified as EDCs has increased to more than one thousand chemicals (Schug et al. 2016), and these can be found in a wide variety of pesticides, plastics, personal care products, and textiles that many people use and consume on a regular basis in their daily lives (Giudice 2016). Regular human interaction with a wide variety of EDCs is at the heart of recently heightened concern about these chemicals, which are being researched by scientists who have departmental affiliations and training in fields from toxicology to obstetrics and gynecology, marine biology to neurochemistry. Most of the DeTox Lab's research has focused on the reproductive and developmental effects associated with exposure to EDCs, even before they were commonly classified under this title. In the *Swimming in Poison* report many of the same EDCs that Zhang studied were referred to by both this term and the less technical "environmental hormone."

In a world where biochemical transformations are increasingly being studied as the anthropogenic effects of pollution, EDCs are often considered particularly worthy of scientific research because, in the words of sociologist Celia Roberts, "they disrupt what are widely perceived as the foundations of life: sexuality, sex and reproduction" (2017, 301). Even if, from an anthropological perspective, one might argue that the perceived foundations of life are not universally consistent, it is true that scientists and environmental activists often describe EDCs as having universal effects, particularly on reproductive and developmental health. Social science and humanities scholars have critiqued this focus of EDC discourse and criticized its heteronormative pathologizing of intersexuality, nonreproductive sexual activity, and impaired fertility. EDC science and activism have been animated by what environmental studies scholar Giovanna Di Chiro (2010) calls "a politics of purity" that focuses on the harm rendered to individualized bodies, particularly their sexual development and reproductive capacity.

Many environmental campaigns at the intersection of reproductive and environmental health, including the campaign behind the *Swimming in Poison* report, also focus on the harmful potential of this chemical class. Like scientists, environmental activists who oppose EDCs often base their opposition on grounds that problematically reify overly simplistic ideas of sex, sexuality, and reproduction, including their primacy. Moreover, EDC research often focuses on the feminizing effects of what are referred to as environmental estrogens, stressing the stakes of pollution for virility and masculinity. Such emphasis places the damage of EDCs on the gendered body, perpetuating gender stereotypes even as it shows the urgency of addressing important questions about the distributed risks of pollution (Langston 2011; C. Roberts 2007). In EDC discourse the chemical threat is often described as a threat to heteronormative order (Kier 2010; Pollock 2016). This concern about sex, sexuality, and reproduction overshadows and even overlooks concerns about other EDC-related health issues, leading to a kind of panic around sex boundaries and sexual deviance (Ah-King and Hayward 2013). But such sex panic is neither a necessary nor universal reaction to increasing concern about the effects of EDCs.

When it was released in 2010, *Swimming in Poison* resulted in an unusual amount of media attention to EDCs or environmental hormones and industrial pollution in China. These media responses, unlike reactions to EDC events in Europe and North America, did not focus on anxieties about sexual purity. Instead, they focused on what some perceive as one of the primary foundations of life in China: food (Farquhar 2002). Furthermore, considerations of food safety and food scandals led many readers of the report and subsequent news coverage to reflect on the broader, ecological scope of polluted fish and to critique industrial capitalism. Poisoned fish were perceived as both embodying humanity's potential future demise and contributing to it. Environmental hormones were consequently thought to have the power to disrupt much more than reproductive and endocrine systems. The disruptive quality of China's environmental hormones had less to do with a puritanical defense of sex or sexuality and more to do with acknowledging the depths to which bodies in China are suffused with the sometimes toxic social, economic, political, and chemical environments in which people eat, grow, and live.

By discussing the particular ways that the threat of EDCs resonated in and around Nanjing and how such concerns went beyond sex panic to create a hormone scandal, might one find the means to imagine a less heteronormative way forward for EDC research? How does a report that was initially inspired by toxicological methods, then went on to inspire toxicologists themselves, who had struggled with reaching audiences beyond the academy, help us rethink the conceptual limits and possibilities of the hormonal environment?

Creating *Swimming in Poison*

At the end of my lunch with Zhang, I scheduled my first visit to the DeTox Lab: a tour that would take place after the university's holiday closure for the Lunar New Year. I took the subway back to my temporary apartment and began researching the Greenpeace report further. I had already planned a trip to Beijing for the holiday to visit other graduate students from the United States who were conducting fieldwork and to meet with research contacts and friends I had met in the summer of 2009 while in China for preliminary research. But perhaps I could also meet with someone from Greenpeace involved with the report, for the organization's holiday closure was likely shorter than the university's break. I reached out to the organization over email and was able to schedule a meeting with someone in the Beijing office to discuss the making of *Swimming in Poison*.[1]

I arrived in Beijing by plane one day before my scheduled interview. The day of the meeting, I took a cab to the address I had been given but spent a few minutes searching in the cold winter weather for the entrance. At the time, Greenpeace China's office was located in an inconspicuous gray building that was difficult to identify from the outside. "Everyone has trouble finding the place," my host, Yu Xiao, said after I had eventually gotten in. She led me through the organization's corridor past meeting rooms and office spaces. We stopped in a small conference room, and after pouring hot water for both of us, she began talking about her work at Greenpeace. She booted up her laptop computer in order to deliver a PowerPoint presentation to just me, an audience of one, as Zhang had the previous week.

During this technological delay we discussed the strategies that Greenpeace uses in its campaigns, which she gleaned from a training camp she had attended in the United Kingdom when she first took the job. She went through a list of tactics: "research, raising awareness, putting pressure on industry." Shortly after stopping, she chimed back in, "Oh, and direct action! Of course, direct action." Though Yu Xiao had nearly forgotten to mention this activist approach, for many, direct action is the foremost strategy associated with the Greenpeace organization, which was founded in 1971 in Vancouver, Canada, as a grassroots organizing body. The idea of direct action comes out of social movements in which activists strive to take power for themselves. Anthropologist David Graeber describes direct action as "distinguished from most other forms of political action such as voting, lobbying, attempting to exert political pressure through industrial action or through the media. All of these activities . . . concede our power to existing institutions which work to prevent us from acting ourselves to change the status quo" (2009, 202). From its beginning, Greenpeace has been oriented toward direct action and is known for its confrontational and creative activism (Lam 2014). Greenpeace today has professionalized offices around the world (Zelko 2013), including several in East Asia.[2]

After Yu Xiao remembered direct action, I asked her if this method of activism was something that Greenpeace China took part in. She responded that although it was important to Greenpeace in general, direct action often seemed inappropriate in their work. I asked what methods did succeed here. She responded that it was hard to know and that campaigns were often learning experiences of stumbling through the unexpected in a country where it was not clear what would work. Her reflections reminded me of common refrains taught to anthropologists about ethnographic methods, in which unexpected findings are said to lead researchers in directions they could not have known before beginning fieldwork. The strategy she discussed was contingent, tied to unclear and inconsistent state restrictions on activism and environmentalism in a country newly coming to terms with its pollution.

In anthropologist Tim Choy's exploration of Hong Kong–based environmental activists, he describes the common absence of reflection on "the cultural specificity of environmental aesthetics and ethics" among environmentalists from the global North (2011, 134). What Choy calls the "situatedness of environmental practice" is often missing as transnational initiatives are put into practice. In other words, there is a lack of attunement not only to the environmental situations that a specific location faces but also to how particular activist practices or the tone of campaigns might resonate, offend, or cause controversy within a specific place. Such attunement was crucial for the development of *Swimming in Poison*. In Yu Xiao's words, the team "followed their intuition" while creating the report, carefully considering what might motivate a Chinese audience to concern themselves with industrial pollution.

For example, the Yangtze River was selected as the focus of the report because of its high historical and cultural value in China, as expressed in songs, art, and television movies. Often called the Mother River, the Yangtze flows nearly four thousand miles. As the longest river in Asia, it starts at the Tibetan Plateau, runs east through middle China, and eventually ends at the East China Sea. The river has played a key role in China's economic development. In 2013 its basin was home to more than four hundred million people (Hollert 2013). The river also hosts more than ten thousand chemical enterprises and takes on enormous amounts of sewage, industrial wastewater, ship navigation waste, and agricultural runoff. Still, the river remains one of the most important sources for the large amount of freshwater fish eaten in China, accounting for about 60 percent of freshwater fisheries production (Floehr et al. 2013).

Similarly, the Greenpeace China team selected fish as a focus of the report because of their symbolic importance—in Chinese, *fish* (*yu*) is a homophone for *plentiful*, connoting abundance and prosperity—and because they are viewed as an important source of nutrition (Oxfeld 2017). *Swimming in Poison* focuses on two regularly consumed fish: the common carp and the catfish. While in our meeting at the Beijing office, Yu Xiao described the cold, wintry trips that she and the team took to the banks of the Yangtze to conduct research for the report as both exciting and difficult. In order to ensure the authenticity of the specimens collected, the group personally visited each of the four sites they had selected along the river: Nanjing, Ma'Anshan, Wuhan, and Chongqing. Yu Xiao and her colleagues recruited local fisherman at fish markets to collect four samples of each fish and watched their boats from the shore as they came in and the fish came off. Fish were purchased for what Yu Xiao described as fair market price and stored on dry ice in insulated containers. The fish were then shipped to Greenpeace's European laboratory for testing.

Yu Xiao continued telling the story of the creation of *Swimming in Poison*, describing how relieved the team was when, after being stalled in Hong Kong by delayed flights caused by the eruption of Iceland's Eyjafjallajökull in April 2010, the fish finally arrived at the Greenpeace Research Laboratory at the University of Exeter. There, scientists conducted tests to decipher the amount of hazardous chemicals in the fish. Multiple toxicologists and other environmental scientists based in China had done similar research on fish from the Yangtze and other bodies of water, furthering Greenpeace's confidence in its pending results and its chosen focus on rivers and fish.[3] The results from the laboratory tests matched what Greenpeace had anticipated. Many toxic chemicals were found in the fish, and the group classified three of these chemicals as environmental hormones.

With the sample collection and results complete, the Greenpeace China team had the firsthand data it needed to create its report. The final product is a well-designed and expertly produced booklet showcasing high-impact photography across many of its pages. The fish data take up limited space, but the particularly Chinese nature of industrial pollution and its national ramifications are stressed throughout the report. As one section reads, "While most countries in the world have greatly reduced the production and use of many of the most hazardous chemicals, largely by means of new legislation, in China both production and use have increased considerably" (Greenpeace 2010, 2). Such internationally comparative contrasts turn pollution and its impacts on fish and their consumers into a national problem. In turn, the solution becomes a matter not of individual action but of government legislation and regulations—at least this is what the Greenpeace China team hoped.

The report also describes numerous health effects of the nation's superlative toxic accumulation. It reiterates that the specific toxic chemicals found in Yangtze River fish have been shown to cause a range of health concerns, but it concentrates mostly on threats to sexual development and reproduction. For example, focusing on the threat of feminization, *Swimming in Poison* describes male fish as particularly at risk: "NP [nonylphenol] and OP [octylphenol] are endocrine disruptors, able to mimic natural estrogen in organisms. This can lead to altered sexual development in some species, most notably the development of female organs in male fish" (Greenpeace 2010, 8). Unlike the choice to focus on one of China's most meaningful rivers or commonly consumed fish, here a concentration on sexual development, feminization, and males "becoming" females replicates an emphasis on sex panic found in much EDC-related science and activism in the United States and Europe.

Critical Approaches to EDCs

In 2003 Celia Roberts suggested that, based on the growing number of scientific journal articles and mass media reports, "It seems we might all be drowning in a sea of estrogens" (2003, 195). Today we have reached a new depth of submergence in EDCs and the discourse that surrounds them. Rates of production and diversification of synthetic chemicals continue to increase (Bernhardt, Rosi, and Gessner 2017). Scientific research on EDCs has exploded since the early nineties—now including many fields and modes of research into the effects of chemicals designated in this class. Similarly, a number of environmental activists have drawn attention to the harmful effects of EDCs since the late twentieth century (Krimsky 2000; C. Roberts 2007; Wylie 2012).

Although the number of EDCs and the amount of scientific, media, and environmentalist attention to such chemicals have grown, the language used to describe EDCs has to a great degree stayed the same. The effects of EDCs are often described in dramatic, panic-inducing language that highlights the potential "disruption" of sex and reproduction, resulting in eye-catching media headlines. As Roberts explains, this language is "dependent on the mobilization of pervasive cultural understandings of sex differences as antagonistic, and of human and other animal existence as based on sexual reproduction" (2003, 202). EDC research regularly interprets increased diversion from "normal" binary sex as a pathology that has an impact on present and future generations. As in other areas of science, EDC research often values sexual activity for its functional, reproductive potential, deemphasizing the potential importance of nonreproductive sexual behavior (Raffles 2010). In EDC studies, queerness is viewed as the aftermath of industrial damage (Pollock 2016).

Environmental studies scholar Giovanna Di Chiro (2010) calls such rhetoric "econormative." On how one might account for the potential harms of EDCs without invoking such sex panic, Di Chiro asks, "Can we imagine environmental-feminist coalitions that can forge a critical normative environmental politics (we *all* should live in a clean environment; we should all have the right to healthy bodies) that resist appeals to normativity?" (203).

In *Swimming in Poison*, Greenpeace China approached EDCs through an econormative lens. By describing sexual difference as harm and stressing the presence and impacts of environmental hormones on men in particular and reproduction in general, the report ascribes to a heteronormative sensibility. Despite the report's econormativity, however, media responses in the wake of its publication interpreted the problem of environmental hormones less as a matter of heteronormative threat and more through concern about food safety, government regulation, industrial capitalism, and ecological harm.

Fishing for Relations

According to Yu Xiao, the goal of Greenpeace's *Swimming in Poison* report was to bring attention to the issue of industrial water pollution and lack of chemical regulations. Greenpeace issued English- and Chinese-language press releases and held a press conference about the report in its Beijing office in August 2010. News of the report went viral in print and social media. Exceeding even Greenpeace's expectations, findings were discussed in more than 115 domestic media outlets, which included 76 print news articles, 10 internet news articles, and 29 op-ed pieces. Moreover, there were thousands of mentions and reposts on the Chinese microblogging website Sina Weibo.

These numbers alone were indicative of a successful campaign. But the quality of the media also impressed the members of Greenpeace China. Many media responses focused on food, diet, and the changing quality of "wild" fish in China's rivers, but some journalists went beyond these details to theorize relationships between environmental hormones and the lack of industrial pollution regulation in a more in-depth manner than anticipated. This was the success that Zhang was so excited about at our lunch in January. Finally, toxicology had been translated to social and print media in a way that resulted in what he regarded as meaningful discussion.

When one analyzes this media coverage more closely, four primary themes recur. First, there was a focus on eating habits. Some journalists suggested strategies to mitigate pollution's effects on an individual scale, anticipating that their audience would reluctantly continue to eat fish from the Yangtze River, even with news of its poisoned state. In response to the report, one editorial suggested, "In addition to enhancing environmental awareness, people should also pay attention to eating habits. Don't eat the same food for every meal, every day, even if you like it, which will avoid the accumulation of hormones, heavy metals, and farm chemicals" (P. Yang, Zhong, and Ge 2010). Besides dietary variation, news reporters also suggested particular cooking methods and tips for selecting safe fish. Such recommendations could point to what many anthropologists describe as the rise of individualism or neoliberalism in China, when individual responsibility becomes the default model for resolving what were once considered collective problems (Ren 2015; Rofel 2007; Yan 2010; Zhu 2013). But such concrete tips for eating in a polluted landscape might also be understood as a means of ethically navigating an overwhelming situation (Tracy 2010). Small acts of food selection and preparation could be interpreted as a means of dealing with a sense of pollution's ubiquity and the unavoidability of contamination.

Despite the many individualized solutions proposed, fresh-caught fish sales were reported to have fallen in some locations directly following the report's publication (Z. Liu, Zhang, and Chen 2010). Media discussions that centered on whether or not to eat fish focused on the increasing difficulty of distinguishing between wild and domestic. In the aftermath of *Swimming in Poison*, reporters from multiple Chinese newspapers debated whether wild fish still provided more desirable food than their farmed counterparts. Such conversations were based on general knowledge that fish from farms are given hormone additives, likely in relation to food scandals revealing that fish had been fed contraceptive pills, as described by anthropologist Yunxiang Yan (2012). Yan points to the emergence of the concept of food safety (*shipin anquan*) in China during the 1990s, a time in which concerns about food shifted from quantity to quality (N. Chen 2010). Around this time, eating became an expression of wealth because of the rise of consumer culture in China, as a growing number of middle-class people had an increasing number of food choices and an increasing number of reasons why certain foods were desirable (Farquhar 2002). This discernment has partially been about the consumption of wild foods (Fearnley 2013). People in China often pay large sums of money for wild foods, including fish. As a journalist at the *Chengdu Business News* (*Chengdu shangbao*) wrote, "Being able to eat purely wild fish is nearly the symbol of luck, status, and wealth" (Xu 2010).

As discussed by Yan, the concept of poisonous food (*youdu shipin*) emerged after the topic of food safety, at the turn of the century, when several food scandals (*shipinmen*) revealed that food had been adulterated for the sake of higher profits. Banned additives were put in feed or food products. Pesticides had been used as food preservatives. Fake food had even been produced from toxic chemicals and water or nonedible substances such as human hair. In Yan's words, "The defining feature of poisonous foods is deliberate contamination" (2012, 710). As such, the prevalence of poisonous food and food scandals in China erodes not only public health but also public trust (Yan 2012). The Greenpeace report cleverly built upon this lack of public trust in food, stretching it into a conversation about "wild" fish in the Yangtze River. Media commentaries followed Greenpeace's lead, refocusing away from the greedy individual who creates poisonous food to pollution that undermines an idea of a pristine wild. In the words of an article from the *Commercial Times,* "The unfounded trust in and pursuit of wild fish has turned out to be the same as eating hormones" (Wu 2010).

Hormones also became the vehicle through which commenters connected fish to humans. The article cited above continued: "When environmental pollution has already been accumulated in the bodies of undomesticated creatures and when environmental hormones have already become pervasive, not only are wild fish poisoned, so are human beings" (Wu 2010). In a more direct example, an editorial in the *Shenyang Daily* (*Shenyang Ribao*) points out that "we, as the upper reaches of the food chain, are inevitably becoming the next 'Yangtze River fish'" (Bi 2010). Here, fish are what historian Brett Walker calls "biological sentinels" or what anthropologist Frédéric Keck and sociologist Andrew Lakoff refer to as "sentinel devices": beings that—through death or obvious ill health—warn others of impending ecological catastrophe (Walker 2011; Keck and Lakoff 2013). But fish are also more than simply warnings; they are nonhuman expressions of the likely state of both the environment and humans.

The final theme arising in news coverage of *Swimming in Poison* was a critique of expert knowledge. Some articles downplayed Greenpeace's findings, quoting environmental scientists who argued that the level of hormones in Yangtze-poisoned fish were low enough that they would not have an impact on human beings who consumed them. Another set of articles was critical of the experts who criticized Greenpeace. For example, as one author wrote, "Several experts have negated the reports of poisoned fish by Greenpeace overnight and issued labels of grandstanding, which is an attitude and tone more like that of governmental officials" (Zhou 2010).

Another writer questioned the expertise of Greenpeace through a different premise—the concerns of ordinary people and their future:

> Ordinary people (*laobaixing*) want to know: even if the amount of hormones is not harmful to the health of human bodies at present, then what about in the future? The pollution of chemical substances in the Yangtze River is not static; qualitative change will happen one day after enough quantitative accumulation, which is what ordinary people worry about. When will this qualitative change that could poison fish and cause

people who eat fish to die occur? Could experts give a schedule for that? (Han 2010)

Other articles sarcastically addressed similar issues. For example, one comment on a news article jokingly questioned the validity of the report: "Greenpeace is fabricating rumors. Rivers, lakes, and the sea have no pollution at all. All water resources have met the standard of drinking directly!" Another comment reads, "It doesn't matter, from poisoned milk powder, melamine, illegal cooking oil, birth control pills, fattening preparations, to antibiotics. . . . We now have strong and sturdy bodies, immune to all poisons" (Xu 2010). Perhaps such humor might be understood as what anthropologist Anna Lora-Wainwright (2017) calls "resigned activism," a kind of quiet political action expressed through resignation. But in this case, instead of expressing resignation, commenters use humor to express the overwhelming extent of toxic exposures. As anthropologist Megan Tracy writes of serious jokes that followed the 2008 melamine milk-powder scandal, "This humour is not simply a political and social critique but also perhaps a survival strategy—how else to live in a world where one's food supplies are a continued source of anxiety and distrust?" (2010, 8).

As these examples show, when *Swimming in Poison* was released in China, the report did not cause "sex panic." The politics of purity at work in the discussion of the report do not emphasize anxieties about the disruption of heteronormative binary sex and sexuality, although they do sometimes question a purification of the wild. Instead, environmental hormones are vehicles through which journalists express concern about the quality of fish as food, as well as the way fish, as a broader part of the food chain and river ecology, indicate greater concerns for humans. This term *environmental hormone*, then, is itself useful as it does at least two things. First, it positions hormones in an intermediary role, as both that which surrounds and that which is internal, thereby encouraging further reflection on the connections between bodies and environments. Second, it connects a concern about environments to concern about other hormonal things, both materially and symbolically, as discussed below.

Eating Hormones

In August 2010, the same month that Greenpeace released *Swimming in Poison*, media attention gathered around another news story, one that might have been viewed as unrelated but was brought into conversation with *Swim-*

ming in Poison through a shared focus on eating hormones. Three infant girls from Hubei Province had been separately diagnosed with female early sexual maturation (*cixing xingzaoshu*) and increased estrogen levels. In order to stop such bodily transformations, the baby girls' doctors suggested that their parents stop feeding them powdered infant formula (Bao Chang 2010; "China Investigates Claims" 2010). According to news sources, soon after the physicians' warning, two sets of parents discovered through informal discussion that they had each given their daughter the same brand of formula, Synutra. Complaints to local authorities and the press followed.

At the request of the parents, and amid media attention to this event, Chinese officials tested the formula and eventually announced that it was not the source of the elevated hormone levels. However, such evidence was difficult to trust. In 2008 six children had died, and more than three hundred thousand were suspected to have become ill from drinking formula that contained melamine, a highly toxic industrial chemical (Tracy 2010). The memory of this tragic event, and numerous other food scandals, justified parental concern about the possibility of what was referred to as a "hormone scandal" (*jisumen*). Even after the official tests disputed a causal association, parents remained convinced that hormones in the milk powder had caused their babies' premature development (Bao Chang 2010; "China Investigates Claims" 2010).

The release of *Swimming in Poison* coincided with news of this hormone scandal. Just weeks later, Greenpeace China held press conferences and distributed press releases on *Swimming in Poison*. Public relations creatively linked the hormonal focus of the milk powder news to the report's focus on environmental hormones, even adding female early sexual maturation to their list of possible effects of exposure in press releases and PowerPoint presentations. EDC exposure *has* been associated with premature puberty (C. Roberts 2015). But by indirectly linking environmental hormones to another hormone scandal, Greenpeace connected the poisoned Yangtze River fish to early-maturing girls in Hubei Province.

Greenpeace China was not alone in drawing such connections; other media were also quick to see a relationship. For example, some reporters and editorial writers tried to make sense of the initial hormone scandal by relating it to the Greenpeace report, asking if fish were the real cause of early sexual maturation or were themselves suffering a similar fate as the infant girls. As an editorial in the *Beijing Business News* joked, "Did these fish eat milk powder when they were growing up?" ("International Greenpeace Organization Publishes Poison Hidden in River" 2010). Another news article

FIGURE 2.1 A cartoon in the *Qianjiang Evening News* depicts Chinese children running from a fish that contains environmental hormones (*huanjing jisu*) (August 31, 2010).

in the *Henan Business News* observed that "the scary thing is that if there are hormones in the Yangtze River, perhaps cows won't be the only organisms that produce hormone milk" (C. Wang 2010).

An editorial from the *QianJiang Evening News* (*Qianjiang wanbao*) goes in a much darker direction: "In 1981 a famous song was written, 'The Yangtze River Song,' which praised the river: 'Your sweet milk has been feeding all national ethnicities' (*minzu*) children.' Is the mother river's 'milk' still so sweet?" (Hong 2010). Next to this editorial is a cartoon of two Chinese children running from a gigantic fish whose body has morphed into a skull, the words *environmental hormone* (*huanjing jisu*) written into a substance dripping from its body and mutating its fins (figure 2.1). The cartoon portrays a sense of children's bodily vulnerability, putting a terrifying spin on feminist science-studies scholar Donna J. Haraway's (2008) idea of multispecies "becoming-with."

By looking at how *Swimming in Poison* was connected with the hormone scandal, one gets a sense of their shared stakes and origins. Both the concerns about early sexual maturation and poisoned Yangtze River fish express a broad sense of social distrust, as well as public concern about food safety, greed, and regulation. Furthermore, both events show anxiety about the role of hormones in animal bodies, both human and nonhuman, and in broader ecological systems. Although undertones of understanding environmental hormones as a threat were present, even these interpretations did not function through an idea of sex panic in the same way as EDC narratives in Europe and North America often operate. This nuanced contrast is important.

Reimagining the Analytic Potential of EDCs

Greenpeace China considered the linkage of the milk powder hormone scandal to *Swimming in Poison* and the resulting media coverage a great success, as did Zhang of the DeTox Lab. For him, the Greenpeace report was an exciting representation of toxicology's potential to meaningfully engage with those living in China. Zhang was hopeful about the campaign's success—a success that was signaled to him by the regional and national media coverage that the report received and the dialogue it had ignited on social media and in person. Indeed, in my own conversations with a medical school administrator at another Nanjing-based university a few days after my lunch with Zhang, the Greenpeace report was brought up again. This administrator's opinion of the report was different—he and his colleagues wondered if the report was flawed, intentionally exaggerating the negative influences of water pollution in the Yangtze River on fish and humans. He introduced me to a group of toxicologists who were designing their own study of Yangtze River fish, one that would likely challenge the Greenpeace report. It seemed clear from these interactions that whether *Swimming in Poison* was interpreted as inaccurate testimony or a model of publicly engaged toxicology, it was evoking attention among toxicologists and others in Nanjing.

Like members of Greenpeace China, Zhang was particularly impressed that within just a few months the campaign reached an even greater accomplishment than widespread media attention—one so far-fetched that it was not anticipated. In January 2011, shortly after the report's publication and the subsequent media avalanche, China's Ministry for Environmental Protection announced that NPs, one of the EDCs highlighted in the report, would be added to the list of toxic chemical substances whose import and export must be regulated through the government. Those who wished to trade NPs would now have to apply for permission and certification. This news came as a surprise to Greenpeace China, as it did to Zhang. *Swimming in Poison* had achieved more than it set out to accomplish, and more than toxicologists had previously accomplished: raising public awareness around industrial water pollution and even contributing to the creation of new environmental regulations.

This success occurred despite media responses to Greenpeace's report lack of focus on the gender-bending, boundary-crossing actions of synthetic, inauthentic chemicals interfering or disturbing the hormones of skin-bound animal bodies. Instead, the media's primary focus was on how the presence of EDCs in China's waterways raises concerns about food safety, as well as the stakes of the ecological relationships between humans, fish, waterways, and the industrial chemicals found in them.

Recently, Celia Roberts (2017) has suggested that although scholars should remain critical of EDC discourse and embrace the queer critique of sexual and reproductive heteronormativity implicit within, they should also be willing to recognize the disproportionate vulnerability of people to the harmful effects of EDCs, based on both their geographic location and socioeconomic differences. This point is important because critical social science research on EDCs increasingly occurs within locations outside Europe and North America or focuses on those locations. To research EDCs in China and other locations is to recognize the historical, political, economic, and cultural factors that influence unequally distributed conditions of toxicity. But it is also important to think through EDCs locatedness in order to acknowledge the varied "cultural nerves" (Ah-King and Hayward 2013, 4) and norms through which EDCs make sense to activists, the media, and the public in various locations and socioeconomic positions.

By studying the creation, distribution, and media responses to *Swimming in Poison*, I have suggested a comparative approach to address the increasing transnational circulation of EDCs, as well as to research EDC science and activism. Certainly, as one of the largest, if not the largest, producers and users of many EDCs, China must be considered within this comparative project. But comparison is not my only goal. Critiques of universality through anthropological conventions of describing "local knowledge" and cultural specificity are not enough (Choy 2011). As suggested by sociologists John Law and Wen-Yuan Lin (2017), as well as anthropologist Mei Zhan (2011), Chinese thought—whether in historical or popular forms—should also be considered for its analytic potential. How might the many responses to Greenpeace's report, published in newspaper, online discussion boards, and microblogs, act not only as a comparative other but also as an analytic inspiration through which scholars might reimagine critical global EDC science and activism?

In the above examples, concerns about EDCs are expressed not through sex panic but through ideas of scandal. This moves the threat of EDCs from a focus on the disruption of sex, sexuality, and a narrow sense of reproduction that provokes anxiety about individual bodily integrity and function toward an emphasis on understanding how scandalous acts result in harm. As more and more research points to a growing acceptance that EDCs should be avoided due to their many negative effects, which appear to go well beyond impacts on sex and sexuality (Ah-King and Hayward 2013), EDC scholarship and activism might stop provoking (or critiquing) panic and start emphasizing the scandals that bring about ubiquitous but variably regulated and unevenly distributed EDC exposures. There is work to be done in reimagining a critical, nonheteronormative environmental politics. Such a reimagining might take inspiration from the scales of responsibilization expressed and circulated through environmental hormones.

THREE

The Dietary Environment

In February 2011 I returned to Nanjing after spending Spring Festival in Beijing, and I began conducting interviews and participant observation in the DeTox Lab. By this time, a lot had changed since the early years of their research. The group had grown from two professors, a few graduate students, and a couple postdoctoral researchers to include nearly twenty scholars at var ious levels. Regional collaborators and affiliations were expanding. Senior lab members were moving up in the university's hierarchy. Graduates were securing positions in research organizations outside the Nanjing Institute of Medicine and Science. For those who stayed on after earning graduate degrees, visits to universities and conferences within and near China were now common. The DeTox Lab's nationally recognized research program was also becoming well known in environmental health programs and institutions around the world.

On my first visit to the lab's facilities after my January meeting with Professor Zhang, I received a tour of the building where lab members worked six days a week, from morning to evening. A graduate student named Wang Bo, whom I would later come to know more closely, met me outside the front entrance, near the many worn-down bicycles standing in rows under a terrace. "This is an old building," Wang told me as we made our way up the steps to the second floor. The statement was implicitly comparative. Many of the surrounding buildings, including the neighboring reproductive endocrinology building, where students learned to treat infertility through reproductive technologies, were newer than the Toxicology Department. Such comparisons, made by members of the DeTox Lab on numerous occasions, highlighted the increasing importance of commercial revenue streams in funding university departments.

Though not as commercially successful as the reproductive endocrinology program, the Toxicology Department had managed to secure enough federal and university funds to soon move to a new, larger space in the institute's neighboring suburban campus. The move was highly anticipated. Zhang had described his current hesitation to invite foreign visitors to his present facility, not wanting to give the wrong impression about the laboratory's capacity. He was looking forward to the forthcoming move to suburban Nanjing, where international scholars and expats often chose to live. Based on my observations in their current space, I imagined how things would change when they moved. Laboratory ceilings would no longer leak. Windows would close securely. Lighting would be improved. Power supplies would be consistent. Many such issues were mundane, but the high stakes of consistent electrical power were made clear one day later in my fieldwork when electricity went out in one wing of the building, leaving specimens in a freezer exposed to higher-than-normal temperatures. Graduate students worked quickly to transfer the vials of blood, semen, and animal tissue to other areas of the building. But in their new space, idealized and imagined, such issues would be a thing of the past, as research infrastructure would beget research capacity (Tousignant 2018).

Most of the DeTox Lab's research to date had focused on endocrine-disrupting chemicals, understood as substances that originate outside animal bodies but enter into them, where they mimic naturally occurring estrogens, resulting in a wide range of influences on health, both reproductive and beyond. Having been found in polar bears and human breast milk, EDCs have global reach. Even those places that are supposed to provide the most comfort—our homes, our clothing, and our favorite foods—have been saturated with EDCs.

These EDCs do not only saturate material goods. They also inundate anxieties about the embodied effects of living in toxic times. These fears find expression through scientific reports, environmental activism, and consumer-advocacy campaigns. They become represented in specific materials: plastic water bottles leaching bisphenol-A (BPA), water polluted with perfluorooctane sulfonic acid (PFOS), and toys covered in lead. The fear of EDCs is like the fear of toxicity more generally; imaginaries of toxicity become affixed to certain chemicals and consumer goods. Because of the "hormonal" frame through which the effects of EDCs are often rendered, anxieties surrounding EDC-specific toxicity are often criticized for being gendered. A focus on how EDCs bring about the feminization of boys and men or the early maturation of girls dominates many, but not all, accounts of EDC science and activism, as discussed in the previous chapter.

But EDC anxiety is also racialized, a point especially important for thinking through the national and transnational dimensions of toxic exposures in China. As queer studies scholar Mel Chen shows, fear of toxic products "made in China," from toys to clothes, textiles to DIY furniture, problematically accelerates the "construction of a 'master toxicity narrative' about Chinese products in general" (2012, 164). Furthermore, Chen argues, the racialization of specific imported products echoes racist fears of Chinese immigrants, a further extension of the perception of a "Yellow Peril" (169).

Such racialized fears of toxicity circulate not only in conversations about the products and by-products of the increasingly synthetic nature of the planet at a moment of transnational industrial capitalism (Casper 2003). They also circulate in discussions of a subcategory of EDCS called "natural endocrine-disrupting chemical" (N-EDC). For example, phytoestrogens are plants, as well as the chemicals contained within plants, that are perceived to have "estrogenic effects" like S-EDCs. Some phytoestrogens are regularly consumed by humans and others by animals.[1] Soybeans and soy products such as soy milk and tofu are the most common phytoestrogens eaten by humans. Unlike the discussion of S-EDCs, analysis of N-EDCs often focuses on both the pros and cons of exposure.[2] Soybeans are thought to have many healthy effects—from cancer prevention to blood pressure reduction, improved immune function to hormone regulation. But they have also been said to potentially "feminize" male bodies and have been studied for their potential link to men's reproductive and developmental health problems.

During my February tour of the DeTox Lab, the first stop Wang Bo and I made was a small room tucked into the northwest corner of the second floor. The room had been set up as a kind of mini-laboratory. As we entered, we saw a junior professor standing in a see-through Plexiglas enclosure that encapsulated both him and a large mass spectrometer. He looked our way and then nodded in acknowledgment of my presence. He climbed out of the enclosure to meet me, closing it behind him. Because the room was mostly occupied by the large machine, we stood close together as we exchanged greetings. He quickly explained his research. Using this mass spectrometer, he said, gesturing to the room around him, he analyzed urine samples to evaluate exposure to toxins via their metabolites. His most recent research had focused on the possible association between male-factor infertility and a diet rich in soy.

At the time of my fieldwork, my first impression of the DeTox Lab's research on soy was that it fell outside the topical boundaries of my study. As I justified focusing my ethnographic attention on pesticides and other EDCs known as toxins, I told myself that soy was beyond my physical and conceptual concerns, exceeding the focus of my project. Soy was not a toxin—at least not in the way that I imagined these things. I, like many others, was caught up in thinking about toxins, and EDCs more specifically, as synthetic chemical pollutants found in a "late industrial" world (Fortun 2012).

But as Hannah Landecker explores in her work on the postindustrial metabolism, increasingly food too is being understood as exposure (Landecker 2011). Landecker describes how the increased permeation of bodies to industrial exposures and their aftermath led to a shift in how twenty-first-century scientists think about and understand metabolism. Landecker suggests that the postindustrial metabolism and the sciences that both study and produce it could "be understood as the biology of risk society" (Landecker 2013, 516). Elsewhere, she writes the following:

> This new metabolism is no longer the interface between Man and Nature, as it was for the nineteenth and twentieth centuries, but a metabolism for the human condition in technical society, where the food is manufactured and designed at the molecular level, the air and the water are full of the by-products of human endeavor and manufactured environments beget different physiologies. This is the character of the study of metabolism in post-industrial nature—the layers of human intervention go all the way down, and the role of biomedicine is to understand and heal the body in the world that humans have made for themselves. (Landecker 2011, 190)

Through Landecker's framework one can see how food becomes exposure and is then researched by toxicologists and others specializing in the study of industrial by-products. In the DeTox Lab's research on soy, the dietary environment is conceptually and materially approached like other kinds of epigenetic or postgenomic environments: with the potential to beget different physiologies that negatively influence reproductive health. In such research, soy becomes exposure. But the postindustrial metabolism is also coupled with the logics of racialized difference, which rebiologizes environmental effects in familiar ways.

Soy as Exposure

The DeTox's Lab study of soy was in part a reaction to research on soy's potential effects on male infertility being discussed in men's reproductive health circles and circulated in international media. Specifically, there was interest in a 2008 study by an interdisciplinary group of Harvard researchers on the relationship between soy intake and semen quality (Chavarro et al. 2008). The Chavarro study contributed human data to a topic that had been extensively investigated in nonhuman animals. Prior research on mice and rats had shown changes ranging from decreased testicular size to a diminished ability to produce sperm and hormonal fluctuations with both direct and in utero exposure to soy, albeit with various degrees of replicability (for example, Adeoya-Osiguwa et al. 2003; Atanassova et al. 2000; Fraser et al. 2006). Chavarro and colleagues worked with staff at the Boston Massachusetts General Hospital Fertility Center to collect semen samples and supplemental data on men's self-reported patterns of soy consumption over the past three months. After analyzing data collected from ninety-nine men, the results of the Chavarro study showed that increased soy consumption was associated with decreased sperm counts.

The study's general conclusion—that higher soy consumption correlated with lower sperm counts—was lacking some nuance when considered from a comparative perspective, as the authors admitted in their discussion of potential limitations. It was difficult to make an argument about soy's universal effects on sperm when men in some areas of the world regularly consumed soy without apparent effect. In order to explain such potential contradiction, authors leaned on an idea of racial difference. Specifically, they suggested that "Asian" populations were biologically distinct from the largely "Caucasian" population that their study researched. Because of the homogeneity of their sample population, the team could offer no statistically significant data to back their claim that the influences of soy were distinguishable by what they called "race." But in order to prop up their suggestion that race was likely contributing to differences in the effects of soy on sperm across populations, they cited another US-based study on reproductive health in "Asian" populations. Conducted in Texas, the study found "Asian" testicular weight to be lower than "Caucasian" and "Hispanic" weights (Johnson et al. 1998). They also cited a World Health Organization report which claimed that "Asian" men (specifically men from Singapore, Bangkok, and four cities in China) had lower testicular volume and sperm concentration than "non-Asian men," albeit at statistically insignificant levels (World Health

Organization Task Force 1996). Through reference to this work, they suggested that perhaps Asian men do experience symptoms related to high soy consumption. Second, they hypothesized that the effect of soy on "Western men" might have more to do with the higher probability that they are on average more overweight or obese than "Asian men," a matter of biological difference tied to dietary environment.

Such hypotheses highlight the problematic use of racial classification in scientific literature in a variety of ways. These include the continued use of race as a biological category, even though it has been shown to have no universal, biological meaning (Brace 2005; Fabian 2010; Fuentes 2015; D. Roberts 2011). Even ethnic categories, which are often based on self-identified cultural and ancestral histories, are problematic because individual research subjects do not define or report ethnicity in the same way. Nevertheless, these ethnic data are then used to compare across studies and across populations. For example, in the Chavarro study, Asian participants in Texas were suggested to be biologically the same as participants in the WHO study who were deemed Asian and who were based in multiple geographic locations, mostly in China. Furthermore, all of these "Asian" research subjects were presumed to have higher rates of soy consumption as well as lack of obesity. Reifying both cultural and biological aspects of an undefined "Asian" identity, the study problematically created an ethnic population through an idea of diet. It then mapped that dietary identity onto an overly simplistic and biologically flawed idea of an "Asian" race. Such haphazard use of racial/ethnic categorizations creates comparative research populations that do more to reify Asian stereotypes than to provide evidence about Asian or Caucasian susceptibility to soy.

The Gender of Soy

Although the Chavarro study encapsulates the continued use of problematic racial categories in scientific research, it more implicitly draws on a long history and contemporary resurgence of comparing "the Asian diet" to "the Western diet." Such comparisons have been made not only through oversimplified ideas of what constitutes "race" but also through stereotypes of sex/gender and masculinity. Soy does have a long history in China, dating back to at least ancient times, when it was considered one of the five staple grains (*wugu*) (Huang 2008). But, as discussed by historian Jia-Chen Fu, soy consumption and particularly soymilk consumption were reimagined and promoted during the Republican Era (1912–1949) as a modern solution to a

Chinese deficit and as a symbol of Chinese growth and development. During this time soymilk not only became associated with "fortification, familial warmth, and protein richness" but also was investigated for its nutritional properties (2018, 5). Public health and educational campaigns viewed soymilk as a key source of nutrition for growing children. Moreover, this scientization of soymilk by Chinese nutritionists was not simply rooted in the past. It adapted dietary recommendations from the United States, where soy and soy products were gaining interest. In the early twentieth century, soymilk was refashioned as a modern path toward nutrition, but one with a long Chinese history that made it particularly suitable for improving the strength of the Chinese individual, body, and nation (Fu 2018).

As the early twentieth century unfolded, soy consumption and plant-based diets more generally became popularly frowned upon in the United States. Historian Melanie DuPuis explains that people on the West Coast, where Chinese immigration and the Chinese Exclusion Act of 1882 had fomented racism and racist portrayals of Chinese and Asian people more generally, white laborers related rice- and plant-based diets to "effeminate and enfeebled" men (2007, 41). Meat and dairy, particularly cow's milk, were more commonly interpreted as the source of "the West's" strength and virility. Lack of milk in the Asian diet was a point through which racial categories were solidified by ideas of racial difference based on habits of food consumption. This occurred despite reports from authors such as Pearl Buck, who argued that the Chinese diet was perfectly sufficient in nutrients, including protein (DuPuis 2007).

In twentieth-century China, soybean products such as soymilk remained central to everyday eating practices, which foregrounded plants until the 1980s. As anthropologist Jun Jing (2000) describes, the consumption of cooking oil, meat, poultry, and eggs nearly doubled in the first decade after reform and reopening. In the late 1990s, dairy milk in particular was promoted as important for children's health, leading to dramatically increased dairy consumption in urban areas in particular during the early 2000s, a growth that can be partially attributed to the dietary recommendations promoted by the Chinese government and the emerging dairy industry (Wiley 2015). From one century to the next, the source of "modern" dietary enrichment was refashioned from soy to dairy. Yet even as dairy consumption rose, many people still questioned the suitability of Chinese bodies to dairy products and wondered if "traditional" ideas of soy were a better biological fit and source of strength for Chinese people.

Such limited trust in dairy as a superior source of nutrition was further compromised by several food scandals that emerged in China in the early

2000s. During my fieldwork in Nanjing in 2011, food scandals were commonly in the news. Toxic chemicals had been found in the featured ingredient of one of Nanjing's local specialties: duck blood soup. Frozen meat distributed through the city's Walmart had been found to contain harmful additives. Fish from the Yangtze River that flowed through the city had been deemed poisoned with environmental hormones. Above all, though, it seemed dairy milk and milk powder were one of the most common sources of food malfeasance and consequent distrust. Following the melamine tragedy of 2008, when six children died and more than three hundred thousand were suspected to have become ill from drinking dairy-based formula that contained this highly toxic industrial chemical, many questioned the increased incorporation of dairy in their family's diet.

Such distrust was partially responsible for a resurgence in soymilk, which I could see around the city. During my own commute to the DeTox Lab and other locations around Nanjing, I would stop at the soymilk stand in the pedestrian underground, which was popular with those on their way to school and work. On the way home I would purchase large cups of milk from another soymilk vendor located in the food stalls and small shops that lined the alley near where I lived. I watched others buy and sell personal soymilk machines from storefront displays outside home-goods shops and within Nanjing's many downtown malls. Soymilk was being promoted on the sides of buses as providing key nutrients *and* encompassing "eastern *yangsheng* wisdom," its traditionally modern quality continuously being reinvented.[3] Making fresh soymilk at home for self and family was now being marketed as a path toward not just a nutritional life, particularly suited to Chinese eating habits and bodies, but also a nourishing life rooted in cultural tradition. This soymilk resurgence was described in the media as a kind of soymilk sentiment (*doujiang aiqing*), an emotional attachment born out of a nostalgia for the scarce and desirable substance from one's youth.

Meanwhile in the United States and Europe, both soy and dairy milk were struggling to maintain consumer faith. Dairy milk was being taken over by alternatives, and soy was a prime contender. But here soy had its own reputational battles. Fears of soy's estrogenic potential was common in popular nutritional discussion, which questioned consumption of soymilk and tofu, and also invoked the fear that genetically modified soy was making its way into many processed food products (N. Chen 2010). Ecological concerns about soy farming rose, as did concern about soy allergies among parents and physicians (for example, Savage et al. 2010). Soy was also a controversial food because of its categorization as an EDC or estrogenic product. Such controversy was solidified through the slur of "soy boys," used by US alt-right and geographically broader men's rights advocates to characterize men who were weak or feminine (Gambert and Linné 2018). In such renderings of (toxic) masculinity, soy and its supposedly feminizing effects became an estrogenic threat that those who consume soy instead of meat are more prone to suffer.

Such ideas of soy's "suitability" for some men's bodies and not others seem to be both expressed in and produced through research such as the Chavarro study. Were the authors of this study suggesting that Asian men's bodies responded to soy differently than do white men's bodies? Or was the Harvard paper arguing that Asian men should have different thresholds through which the effects of soy are measured because of biological differences in men's reproductive parts and substances? It was unclear to a number of critical readers, including those in the DeTox Lab.

Interindividual Differentiation

Senior members of the DeTox Lab, who had been studying the effects of EDCs on sperm for many years, felt compelled to respond to the Chavarro study. If soy was truly contributing to the decline of human sperm, then surely their research on male infertility in an "Asian" population could find it. Partially in response to the Harvard article and partially out of curiosity about whether this N-EDC was influencing sperm in ways similar to the S-EDCs they usually researched, the DeTox Lab designed a study to see if high rates of soy consumption influenced sperm quantity or quality among the "Asian" population in Nanjing.

Working through established hospital affiliations, the DeTox Lab quickly recruited more than a thousand men who had been diagnosed with male infertility or (for the control group) had successfully become fathers in the preceding six months. This sample population was more than ten times the size of the Harvard study, promising a more robust account of soy's spermatic effects. Such recruitment was possible because of a number of factors, including the increasingly common treatment of men's sexual and reproductive health issues in China through men's medicine (*nanke*) and reproductive technologies (E. Y. Zhang 2015; Wahlberg 2018b). The DeTox Lab assumed that as a result of what it called a "traditional Chinese diet," all recruited men would have soy-consumption levels on par or higher than the men in the Chavarro study, a factor that could be measured via soy metabolites. Through a comparative approach, their study asked how the dietary environment, measured through soy metabolites, might affect Chinese men differently than their Western counterparts.

The lab's research echoed the problematic approach to race/ethnicity put forward in the Chavarro study. This is not to say that the DeTox Lab's research had previously excluded considerations of race. Previous publications conducted by the lab had reported that 100 percent of their research subjects' race was Han Chinese. But up to this point, race had been what might be interpreted as an empty demographic—a statistic that, like participation-consent rates and human subject clauses, qualified their articles for consideration in English-language publications based abroad but did little more. One member of the DeTox Lab told me that the "race" data used in their research came out of the demographic survey collected by the affiliated hospitals from which their samples came. But it was unclear to him, and to me, whether race was reported by patients or judged by those administering surveys. However, it was clear that the category of race itself had been a largely empty signifier for the lab to date, stripped of meaning through its application to every person from whom samples were collected.

But when it came to studies of soy, the DeTox Lab was newly immersed in a specialty area of research that had explicitly been racialized by those who their study would reference. Race was newly essential as Chinese men's ability to metabolize soy came to the fore in the DeTox Lab's study of the dietary environment's influence on sperm health. Was the Chinese population, measured through the study of a thousand men in Nanjing, more adept at soy metabolization? Were "Chinese sperm" immune to the effects of soy? As the lab researchers began their analysis through such questions, they found that men with higher levels of certain soy metabolites were more likely to be diagnosed with unexplained infertility, as well as several semen abnormalities, including low sperm counts and low motility.

This finding seemingly confirmed the results of the Harvard study. However, a wrinkle appeared in their analysis that provided more nuance to their results. The men who had been used as "fertile controls"—who had helped conceive children within the past year—also had high rates of "phytoestrogen exposure," but they had not suffered harmful effects. These findings suggested that Chinese men in general were not immune to the negative influences of soy but that certain men consumed high levels of soy without changes to semen quality. The lab's question then shifted from Does soy consumption negatively influence Chinese men's semen quality? to Why does soy consumption negatively influence some Chinese men and not others? In other words, what sort of difference made a difference?

The DeTox Lab researchers began brainstorming factors that might be contributing to such different effects of phytoestrogens among Chinese men, eventually finding themselves at the idea of genetically differentiated metabolisms. Specifically, the team hypothesized that certain single nucleotide polymorphisms (SNPs) were underlying interindividual metabolic difference. SNPs are defined as single-base-pair changes in the sequence of DNA.[4] As discussed in the work of Duana Fullwiley, after the Human Genome Project SNPs became "the primary units of comparative genomics" (2007, 4). Following the work of Troy Duster (2005), Fullwiley argues that even though the Human Genome Project found 99.9 percent genetic sameness among humans, such findings did little to undermine the biologicalization of race or racial categories. Instead, the vast amount of genetic sameness was replaced by fractional amounts of genetic difference, which were scaled up to become the new physical terrain of sciences that continue to genetically mark race. As Fullwiley explains, "Where most people might have a stretch of ATGCCTTA in a genetic sequence, a few might have ATGCCTTG, or a single polymorphic change. Such potential polymorphisms were observed to happen at about every 1,000 base pairs, or DNA letters. Analysed alone, or taken together in blocks known as haplotypes, they are now the primary units of comparative genomics" (2007, 4).

The DeTox Lab used SNPs in exactly this way. From a literature review, members of the lab identified SNPs commonly thought to reduce the activity of enzymes involved in the metabolizing of phytoestrogens. If these identified SNPs were more commonly found in a subset of Chinese men, this finding might point to what made some more capable of processing soy than others within and outside of China. In particular, the lab compared SNPs commonly found in Han Chinese men from Beijing to those commonly found in men with Western ancestry living in Utah. These population subsets in Beijing and Utah were then made to stand in for Han Chinese and US Caucasians in general, which became the comparative units derived from SNPs. Through this comparative expansion, "Chinese men" were shown to more likely be able to metabolize soy phytoestrogens at a healthy rate because of "genetic difference."

Racializing the Metabolism

When male-factor infertility research became focused on the dietary environment, discussions of race/ethnicity became explicit, and the presentation of racial data became meaningful in a way it had not been before. Previously merely a box to check in the biographic descriptors generically provided about human research subjects involved in infertility research, in studies of soy and sperm race/ethnicity became the most relevant lens through which susceptibility was understood. Racialized ethnic categorizations became essential and essentialized components of a familiar geographic comparison between "Asians" and "Caucasians." In the DeTox Lab's study of the interindividual variability of soy's effects, an idea of what it means to be Chinese comes into being. This idea of Chineseness is about both biology (SNPs) and culture (dietary customs and traditions). Here, the reification of race/ethnicity occurs through a genetic understanding of what makes metabolic difference.

Becky Mansfield (2012) makes a similar point about how epigenetics studies that see food as an environmental exposure can reinscribe ideas of race through both the oversimplifications of dietary norms and the characterization of risk. Mansfield argues that "an epigenetic understanding of life does not eliminate but rather transforms notions of biological race and can intensify racialization, simultaneously producing race as a category and ascribing causality for phenomena to racial difference" (2012, 353). Similarly, Natali Valdez describes how epigenetic studies and forms of postgenomic research face various methodological constraints when trying to account for the roles that race or racism play in environmental exposures (2021). Such research shows that even the biocultural explanations of causality that come about through studies of what Landecker (2013) calls the "new metabolism" are often accompanied by old forms of thinking about physiology as destiny. In this sense, metabolic thinking is more than an epochal historical development that turns everything—plastics, pesticides, foods—into exposures. Metabolic thinking is also, as anthropologist Harris Solomon shows, a project that works to find and biologize difference in unfortunately familiar ways (2016).

In the DeTox Lab's studies soy is not just exposure. It is an exposure with different stakes for differently classified bodies: men and women, Asian and Caucasian. Through reasserting an idea of Chinese race and traditional culture as distinct racial and ethnic categories, then ascribing causality to such biocultural difference, the lab reasserted a familiarly deterministic idea of race as genetically definable, even if interindividually variable, resulting in the racialization of the metabolism.

FOUR

The Maternal Environment

The walk from the Nanjing Institute of Medicine and Science's downtown campus to the affiliated children's hospital is not far. Wang Bo, a graduate student at the DeTox Lab, and I have plenty of time to stop at a fruit vendor's stand to purchase a watermelon as a gift for the neonatal surgical unit staff, who have agreed to meet with us today. We will be meeting primarily with a medical student named Lin Ming to discuss his work treating infants with a variety of conditions requiring neonatal surgery. Lin and I have met briefly before at the DeTox Lab. There, Lin prepares and analyzes sample for studies on the potential epigenetic factors involved in the development of congenital disorders.

As Wang and I enter the hospital's pavilion, watermelon in hand, we maneuver through the crowds of parents and grandparents gathered around sidewalk vendors selling balloons and plastic toys for the children being treated inside. Upon entering the hospital's main lobby, I am struck by the number of family members who accompany each child. Together, multiple generations sit on every available chair and bench, stand at every corner, and sit on every window ledge of the large room. We board the elevator, where children and infants with bandages and face masks sit on the hips of their caretakers, meeting us at eye level. This is a premier children's treatment center for not only Nanjing residents but also those living in the surrounding areas of Jiangsu and Anhui provinces. Indeed, many of the families at the neonatal surgical unit are from far away. While infants sleep in small mobile cribs, parents and grandparents rest beside them on cots or the floor so that they will not have to leave their children or pay to stay at hotels nearby. As I will soon find out from Lin, parents of infants who suffer from congenital disorders tend to blame themselves for their child's ill health, something that he hopes to discourage through both his research and patient interactions.

Physicians in training such as Lin try to discourage parental guilt and embarrassment surrounding congenital disorders, pointing to the many possible causes of the conditions they treat. These causes go beyond individualized behaviors, pointing to air and water pollution, food safety and availability, and changing occupational conditions that have accompanied broader shifts in Chinese daily life during the past thirty years of social and economic change. In his regular interactions with the parents of infants he helps treat, Lin suggests that many factors—few of which are the parents' responsibility—could have brought about such disorders. Such a rendering of causality may come across as an anecdotal remedy offered to parents in need at a stressful postnatal moment, but the hospital had recently partnered with the DeTox Lab to investigate the potential causes of congenital disorders. This is how Lin ended up working at the DeTox Lab, where he often prepares and processes samples collected at the hospital.

The DeTox Lab's interest in how indirect exposures to toxins may be linked to congenital disorders and their potential intergenerational transmission comes out of concern about the high rates of congenital disorders in China. China's Ministry of Health has reported that "birth defect cases" increased by 70 percent between 1996 and 2010 (Ministry of Health 2011). According to anthropologist Susan Greenhalgh, this "sharp rise in the number of 'defective infants' has led to a renewed concern about the biological 'quality' of the next generation" (2009, 215). In China, ideas about the burden of parental responsibility, especially maternal responsibility, for fetal health have been exacerbated in recent decades by government pressure and familial expectations to reproduce children who can contribute to a high "population quality" (*renkou suzhi*). Government campaigns aimed at encouraging parents to focus limited resources on fewer children, as well as prenatal health programs meant to increase fetal health and thereby population quality, continue to affect Chinese families through the paradigm of "informed choice" (Zhu and Dong 2018). Blame for congenital disorders or "birth defects" in particular often falls on mothers, interpreted as a failure to fulfill the burden placed on women to safeguard fetal health by, for instance, undergoing prenatal genetic testing or wearing radiation protection vests during pregnancy (Zhu 2013; J. Li 2020).

In epigenetic research on congenital disorders, such patterns of maternal blame have been recapitulated through the idea of the "maternal environment." Social scientists have rightfully criticized such conceptualizations, showing how such research exaggerates preexisting ideas of maternal responsibility for child health (Lappé 2016; MacKendrick 2014; Richardson et al. 2014; Valdez 2018; Warin et al. 2012). Researchers have also showed that even when understood as not-equivalent entities, the maternal environment is often mapped onto individual mothers and their behaviors (Pickersgill et al. 2013). Such critiques of the maternal environment are important because they point out that even in research paradigms that stretch beyond the body to the environment, gendered reproductive burdens are reiterated.

In conversations with Lin and Wang during my fieldwork, I found that the DeTox Lab's research on congenital disorders had the potential to increase such maternal blame for disease by reducing the mother to a maternal environment. But their studies also encouraged a reimagining of disease responsibility as existing in relations and configurations that move beyond individualized notions of maternal personhood. In the DeTox Lab's collaborative studies on congenital disorders, the maternal environment is not only the in utero influences of an individual. The maternal environment is also a lineage of exposures, a series of people embedded and embodied in the sometimes-toxic human and nonhuman environments of the past and the present. In this environmental instantiation, epigenetic research deindividualizes ideas of the maternal environment and of disease responsibility in and through the study of intergenerational, epigenetic inheritance.

What If the Environment Is a Person?

One of the first anthropologists to discuss epigenetics was Marilyn Strathern (1991, 585–88), who proposed at least two alternatives to what she describes as the Euro-American folk model of individual and society "replicated and miniaturized" within genetic and epigenetic research. First, she provided a cross-cultural comparison, describing how Melanesians conceive of themselves not as independent persons but as persons always in relation to other persons, or "dividuals." Second, she brought this dividual back to a discussion of epigenetics. After listening to a presentation on the revolutionary potential of epigenetics at the 1990 meeting of the British Association for the Advancement of Science, she followed the lead of a gynecologist also in attendance to ask whether the epigenetic environment with which the gene interacts could be another person—for instance, the mother hosting the impressionable embryo.

Indeed, Strathern and her gynecologist peer were on to something. Since 1990, research on what is now called the "maternal environment" has become common in developmental and reproductive toxicology, as well as a number of other disciplines that are looking into the intergenerational effects of environmental exposures. Social scientists have described epigenetics in general as a new iteration of a long-standing pattern that blames mothers for aberrations in fetal or infant health. Epigenetic theories of disease causality can reinforce these patterns of maternal blame by turning individual maternal behaviors or lifestyle choices into toxic exposures that impact fetal health (Mansfield 2012). For example, epigenetic research and its media representation in United States and Australia have depicted the responsibility for epigenetically inherited conditions as the product of individual maternal behaviors such as smoking (Warin et al. 2012) or of essentialized behavioral patterns such as diet (Mansfield and Guthman 2015). Epigenetic research that draws attention to, and only to, the mother as the source of future disease causality is certainly limited by assumptions of individuality underlying such research.

This logic of maternal blame relies on an assumption of two individuals—a mother, depicted as a badly behaving vessel, and a developing fetus, the victim of the mother's exposures and behaviors. Working through an emphasis on the uterine environment, the mother is figured as the primary site of "epigenetic becoming" (Mansfield and Guthman 2015, 6), where and when infants simultaneously archive the past and become the future. The mother, whose diet or behavior has affected the fetus, is plucked out of any type of context—social, environmental, national, chemical—and turned into an individually responsible being, stripped of her surroundings (Landecker 2014). As stated by Martyn Pickersgill and colleagues, "In effect, women are framed through some kinds of epigenetics research *as* the first environment for children, potentially activating and augmenting a range of moral discourses and subjecting [women] to (increased) scrutiny" (Pickersgill et al. 2013, 437).

In conversations and observations at the DeTox Lab and its affiliated hospitals, however, the maternal environment resonated more with what many anthropologists have interpreted as organizing principles of everyday Chinese life than the individualization and responsibilization that is said to characterize biomedicine in Euro-American contexts (Clarke et al. 2003;

Gordon 1988; Scheper-Hughes and Lock 1987). These organizing principles include a more relational notion of personhood (M. M. Yang 1994), a correlation between macrocosm and microcosm (Sivin 1987), and an analytic of oneness between humans and environments (Tu 2001; Zhan 2011). In the laboratory where I conducted research, relationality was newly configured through epigenetic research as a response to changing economic, social, and environmental conditions. Instead of, or at least in addition to, reinforcing an idea of individualized maternal responsibility, the lab's research highlighted the connections between humans across generations, within and as environments, thereby reasserting a sense of intergenerational connectivity and responsibility.

At the Children's Hospital

Wang Bo and I enter the neonatal surgery unit and present the children's hospital staff with our gift, the watermelon. After quick introductions we put on white overcoats to begin our tour of the unit. In the hospital setting where Lin spends most of his time, his quiet laboratory demeanor becomes that of a confident physician in training. Lin is clearly in his element; Wang and I are clearly out of ours. The lead nurse joins us as we move through the five or so patient rooms. In each room there are between three and five infants, accompanied by parents and often grandparents. Between nine and fifteen people occupy each space—some adults sleep, some sit on makeshift hospital beds, and some cradle their babies who are either awaiting or recovering from surgery. All are there for the duration of their infant's stay, only stepping outside and into the Nanjing city streets for food or a bit of exercise.

As we move from patient to patient, parents and grandparents assist in removing blankets and surgical dressings to allow us to view the infants' bodies and fleshy traces of operations. Parents and grandparents watch me even more carefully than I am watching Lin and the head nurse. I do my best to remain stoic, afraid I might express my sadness and fear. I save my questions for afterward and nod as I learn about each infant: their condition, whether or not surgeries have been successful, and when they will most likely be discharged. One child has recently had a mass removed from her head. Another was born without an anus. Many are at different stages of recovery after having been surgically treated for rectal or intestinal deformities, the most common conditions diagnosed in the unit.

After our patient visit, we wash our hands and sit in a small back room where the lead nurse serves us the watermelon that Wang Bo and I have brought for the staff. The prearranged interview feels more like a conversation between colleagues; Wang Bo is equally interested in the discussion with Lin. Wang has been researching the epigenetic mechanisms of reproductive diseases for years now, but unlike Lin he does not interact with patients. Since he spent a summer working in the toxicology laboratory as an undergraduate, his plan to follow in his mother's footsteps and become a doctor changed. He has now worked as a research scientist for more than five years, and in all that time this is his first trip to the children's hospital, where some of the tissue and blood samples used in the laboratory are gathered. His curiosity catches me off guard; his excited inquisitiveness seems to surprise even him. He is brimming with questions about patients, parents, and procedures. What conditions do they face? What treatments do they receive? How often do they survive? He does not ask what causes the conditions because, like Lin, he knows that the answer to this question is multifaceted and is at the heart of their collaborative study.

Lin Ming contributes to the DeTox Lab's epigenetic research on congenital disorders, but he spends most of his time treating those patients in need of neonatal surgery. Indeed, it was through conversations with parents and grandparents accompanying neonatal patients that Lin developed both his commitment to becoming a doctor and his perspective on the importance of conducting research into the causal factors of birth defects, he tells me. Lin explains that patient and parent interactions are what led him to truly commit to his work as a doctor. Being a doctor in China is difficult, he maintains, so his level of commitment to his work is clear. Lin tells me that he is unsure if his coworkers share the same feeling about the profession.

The commitment he described was illustrated through a story he told of a night when he was on call. A couple came to the hospital with their newborn son. Desperate for some kind of answer to their child's inability to eat, sleep, digest, or pass a bowel movement, they had traveled from the Anhui countryside to Nanjing. They pleaded with Lin to save their child's life. After hours of examinations and tests, Lin was finally able to diagnose the child with an intestinal abnormality and rush him to surgery, where physicians performed a simple procedure. He was proud of his accomplishment, of his correct diagnosis. But he said the moment that truly changed him was when he went to tell the parents the news. On hearing about the successful surgery, both parents, crying, fell at his feet in what Lin interpreted as the ultimate symbol of gratitude.

While he told me this story, Lin himself was moved. He continued reflecting: this kind of interaction made him realize he was happy to be a doctor. Regardless of the long hours and frustrating patient interactions that accompany his career, the profession brings him great joy. The elation that comes from helping others is something that he hopes his generation of Chinese students and professionals will also find. Yet in actuality, he thinks his peers are much more concerned with making money and buying a house (practically a requirement for anyone hoping to marry, Wang adds from the other side of the table).

After a pause, Lin second-guesses himself, giving his peers the benefit of the doubt. Many of them are from the country, he says, so the idea of buying a house in the city is so far-fetched that they believe that they must put all their energy into such material pursuits. For Lin, the ethical expectations of today's youth in China must be couched within an understanding of the importance of geography and opportunity. Geographic separations, which largely represent economic divisions as well, keep certain graduate students from pursuing careers for more than the financial advantages they bring, which is seen as their only hope for fulfilling familial duties, Lin explained. Wang agreed: "Without a career there is no money; without money there is no marriage; without marriage there is no child."

In Lin's description of his peers, one can sense both a critique of increasing materialism among young adults in China and an understanding that such focus on desiring and acquiring results from broader environments and trends (Ren 2013; Rofel 2007). Socioeconomic circumstances, gendered kinship obligations, and familial expectations greatly shape career trajectories and professional motivations, Wang and Lin agreed. This second understanding of subject formation, as always occurring in relation to other human and nonhuman environments, also exists within the DeTox Lab's research on congenital disorders, particularly in its study of Hirschsprung's disease.

Researching Hirschsprung's Disease

At the time of my hospital visit, Lin Ming was assisting with a project investigating the potential heritability of Hirschsprung's disease through a study of the epigenetic mechanisms involved in its development. Hirschsprung's disease (HD) is often described as a disconnected communication signal or power line in the intestinal tract. Infants diagnosed with HD lack nerve cells in their intestines and are unable to "send messages" or "transmit signals" that should move fecal matter through the colon and out the rectum. This leads to constipation, which can then lead to a distended abdomen, malnutrition, infection, and even death. The most reliable symptom of HD is a lack of bowel movement in the first forty-eight hours of life. Yet many infants with the disease do have occasional bowel movements, often making the condition difficult to diagnose. The disease can be life threatening if left undiagnosed or untreated, and it is said to affect one in every five thousand births worldwide (US National Institute of Health 2013). It has been argued that these statistics vary by "race"/ethnicity; a Taiwan-based study claims that for every two Caucasians with HD, 2.8 Asians were found to have it (Chin and Chiu 2008). National concerns about reproducing healthy children and a quality population thus likely informed the DeTox Lab's decision to focus on this particular condition, as did the fact that males are four times more likely to be diagnosed with HD than females (Emison et al. 2005).

To conduct research on the possible epigenetic mechanisms involved in HD's pathogenesis, the DeTox Lab partnered with a children's hospital in Nanjing that treats about seventy-five HD-diagnosed infants per year. Lin seemed to regard his time working at the hospital as difficult and exhausting, yet on another day when I saw him analyzing specimens in the DeTox Lab, he described the laboratory aspect of his work as a painful hardship (*xinku*), with long hours and much repetition. As the primary technician on one of the lab's investigations into HD, Lin turned his attention to the analysis of samples that may shed light on the epigenetic mechanisms involved in HD.

Epigenetic mechanisms are now thought to regulate gene expression, a process that occurs through a variety of activities, including DNA methylation (the addition of methyl groups—a group of atoms, three hydrogens and one carbon—to DNA). Methylations block the transcription of DNA to RNA, a necessary process in gene regulation, but a process also associated with some health conditions, such as HD. By investigating methylation patterns, the DeTox Lab analyzed whether genes thought to be related to the production of neural crest cells (which migrate to the intestines) have been deactivated via methylation. The lab found that genes involved in the formation of critical colonic nerve cells are turned off by specific methylations. Once turned off or deactivated, intestinal nerves are less likely to develop, and bowel motility is reduced—a conclusion reached by comparing the methylation patterns derived from the analysis of colon tissue and blood samples from patients diagnosed with HD to a control population's samples. This research investigates whether HD, and potentially other congenital disorders found in infants in China and around the world, might be epigenetically inherited. It is a first step toward studies of HD that will not only pinpoint the mechanisms through which epigenetic transmission occurs but will also indicate environmental exposures that bring about such epigenetic modifications in the first place.

To study the toxic exposures that lead to modifications such as DNA methylation, the lab has also designed a multiprovince birth-cohort study. Birth-cohort studies are often considered ethical to conduct on human populations because they are so-called natural experiments. They often claim not to create, but instead monitor, the activities, behaviors, or physical/social/psychological attributes of their research subjects. Such studies generally track individuals over the life course and across generations, claiming to observe research subjects rather than to affect them through exposure to certain influences or research variables. In epigenetic research, birth-cohort studies are used to investigate the potential intergenerational impact of epigenetic inheritance in humans, both by monitoring the biological samples of multiple generations over time and by collecting detailed information about the environments in which people work and live.

In the DeTox Lab's research on HD, samples and their accompanying environmental characteristics have been gathered from multiple generations: infants, parents, and grandparents. Blood, urine, and semen samples that will eventually be used in studies of HD and other conditions come from patients at university-affiliated hospitals in the Yangtze River Delta provinces and beyond. When these biological samples are taken, surveys simultaneously collect information regarding occupation, residence location, and potential toxic exposures. Additionally, researchers often gather general air and water pollution figures for the surrounding areas. Lab director Zhang Zhiyuan's hopes that this broad collection of samples will eventually allow him to correlate methylation patterns now known to be involved in the development of HD with particular environmental exposures. Here, environments are understood as exposures (Landecker 2011), in a variety of human and nonhuman forms. Data are collected in a number of ways—biological samples, written surveys, and chemical analyses—and these various versions of the environment are considered in and through one another.

At the time of my research, publications had yet to result from this sample and data gathering. But researchers hoped that in the future, statistically significant patterns between patients with HD and past generations' exposures to specific environments would be found. Such epigenetic inheritance requires the understanding of environment as person: the mother becomes a vessel through which exposures and their associated methylations travel. But it also encourages an understanding of the epigenetic environment interacting with genes as more than simply an individual. This research places the mother of an infant who suffers from HD in relation to multiple human and nonhuman environments. Even when the environment is a person (the mother), this person will be seen as a part of many additional environments that make up and surround her, both personal and impersonal. These could be human relatives from previous generations, as well as industrial, water, or air pollutants; pesticides; and/or other occupational exposures. Through these multiple environments the DeTox Lab's research draws attention to the toxicity of China's recent industrial, economic, and social changes on many scales, going both back in time and outward from the interior genetic structure to the surroundings of the fetus.

Epigenetic In/dividualism

The interpretation that women actually become the environment in epigenetic research harkens back to the questions posed by Strathern and the gynecologist she was inspired by at the 1990 meeting. However, instead of considering such framing of woman as an environment problematic, Strathern saw the person-environment collapse occurring in some epigenetic studies as a potential means to challenge the very concept of the individual at the heart of so much genetic thinking. This collapse, where and when the maternal environment comes into being, also has the potential to disrupt the individualism at the heart of notions of "epigenetic responsibility" (Hedlund 2011). In the DeTox Lab's HD research, the cause and solution to epigenetically linked problems were not narrowed to the individual (Warin et al. 2012, 370) or to the spatiotemporal coordinates of the uterine environment (Guthman and Mansfield 2013). Instead, proposed birth-cohort studies, physician interactions with patients, and a focus on toxins that result from industrialization recontextualize the fetus in time and space, placing "it back into the uterus, and the uterus back into the woman's body, and her body back into its social space" (Petchesky 1987, 287). In terms more often used in characterizations of Chinese settings, the fetus is placed at the center of concentric circles of interaction.

Concentric circles have often been used in classical Chinese texts to describe the multiple scales of human relationships, where the self lies at the center, followed by the family, community, country, and, eventually, all of humanity at the outer limit (Tu 2001). Sociologist Fei Xiaotong's famous description of Chinese kinship similarly likens interpersonal relations "to the concentric circles formed when a stone is thrown into a lake" (Fei 1992, 63).

Anthropologists often characterize China and East Asia as places that think of bodies as similarly never existing in isolation (N. Chen 2003; Kleinman 1981; Lock and Farquhar 2007; Scheper-Hughes and Lock 1987). Instead, anthropologists emphasize a long history of diagnosing and treating the social causations of bodily illness through an understanding of coconstituted human-environments (Tu 2001; Zhan 2011). Such relational ideas of personhood have characterized social scientific models of Chinese life and are similar to those found in the DeTox Lab's epigenetic research.

This assertion does not mean to explain the DeTox Lab's approach to epigenetic research as simply a product of Chinese thinking or to assert that there exists a singular or timeless Chinese understanding of the body, self, or personhood as relational. The concentric circles that emerge from a stone dropped in water are not stable. They shift and move with the water they are in, dynamic and at times overlapping. That being so, perhaps as DeTox Lab toxicologists articulate the concentric circles of human and nonhuman kin as epigenetic environments at various scales they are to some extent drawing on long-standing relational modes of understanding social life related to the interconnection of humans and environments in order to criticize more contemporary disconnections in China. I interpret the DeTox Lab's research not as an extension of traditional Chinese thought but as an effort to reemphasize connections between inner and outer environments, self and kin, human and nuture that once figured more prominently in daily life. At the very least, their recontextualization of the fetus into concentric circles of multiple human and nonhuman environments is what medical student Lin hopes to communicate to the families of his hospital patients.

According to Lin, families who travel to the children's hospital from rural areas of Jiangsu or surrounding provinces are often embarrassed by their infant's deformities, especially when the infant has no anus. Lin believes that this shame derives from the common belief among rural parents that their children's congenital disorders are karmic retribution for either their bad deeds in past lives or their lack of filial piety in the present. As a physician, Lin hopes that his research on the potential epigenetic mechanisms involved in the transgenerational inheritance of birth defects will encourage parents to stop blaming themselves for their children's conditions. He and other members of the DeTox Lab conduct epigenetic research to offer a response to China's rapid social, economic, and environmental change by understanding how and which toxic exposures are epigenetically inherited. Here, to understand the environment as a person is to understand the mother as always in relation to other environments and people—and therefore able to take only partial responsibility for the health of present and future generations.

Lin hopes such displacement, both from the present to previous generations and from parents' traditional beliefs to the science he practices, will ease the sense of responsibility that burdens so many parents. Likely, Lin paints too simplistic a distance between his scientific understanding of HD and parent interpretations of their infant's condition. Yet Lin's thoughts on this issue offer a nuanced contrast to critiques of epigenetic sciences that claim such research furthers individual, maternal responsibility for fetal health. In Lin's research and treatment practices, the epigenetic environment is not mapped onto an individual mother's body but instead extended and multiplied through a more generous understanding of the maternal environment; it is made to speak back to feelings of responsibility expressed by his patients' parents. Such research and treatment practices show that an epigenetic understanding of disease causality can be, in some instances, a means of thinking beyond the individual.

Lineages of Exposure

In some ways, the DeTox Lab's research contrasts with epigenetic approaches that strictly focus on maternal environments. The lab's research emphasizes an idea of persons as partial entities that exist across generations and intertwine with surrounding human and nonhuman environments. This contrast proves important because the kind of persons and environments brought into existence through epigenetic research have the potential to be translated by patients, the media, and even policy makers into ideas about who or what is responsible for exposures. At the lab, this means responsibility for the conditions of toxicity potentially causing epigenetic alterations. By collecting multigenerational and multivariant evidence for the intergenerational transmission of epigenetic transformations, the DeTox Lab works through an understanding of persons as dividuals, thereby emphasizing a responsibility for toxic environments that goes beyond the mother, beyond her relatives, and into aspects of the physical, chemical, and sociocultural environments that surround and compose toxic lineages of exposure. These are environments that do not simply exist as stagnant surroundings but have been shaped and formed through the past thirty-five years of China's social and economic change and industrial development. In the case of the DeTox Lab, an interdependent understanding of person-environment relatedness challenges the linear and individualistic trajectories of both a strict Mendelian genetics and a changing China after reform and opening.

The DeTox Lab's hope for epigenetic research on congenital disorders and reproductive health conditions is not that it will change maternal behaviors, lifestyle habits, or even personal exposure levels. Instead, the scientists pursue research to capture a much more complex picture of exposure, a more varied and elaborate toxicity linked to industrial development, one that they fear has been and will continue to be inherited. Their epigenetic research constitutes a means through which a small group of scientists critical of the conditions of toxicity that surround them can provide evidence for what they see as the harmful nature of relentless economic development and lax industrial regulations. Furthermore, it is a scientific project figured in embodied reproductive consequences. Materialized in blood and bowel, epigenetic biomarkers of disease become a means for the DeTox Lab to relay to Chinese public health officials, and others open to listening, that the quality of China's future, and of future generations, is at stake.

FIVE

The Laboratory Environment

During my fieldwork in Nanjing, Professor Zhang Zhiyuan traveled to the United States to visit laboratories conducting research similar to his own. Over six weeks, Zhang presented his most recent studies of EDCs in China and worked to advance US research collaborations with various venues. During Zhang's extended trip to the United States I spent much of my time at the DeTox Lab with Wang Bo. As I spent more time with Wang, I came to understand his research, which focused on distilling the precise epigenetic mechanisms involved in the intergenerational inheritance of male infertility through animal models.

Wang's approach was one that might be classified as "environmental epigenetics," which is often defined as the study of how environmental exposures induce epigenetic changes, particularly during "critical development windows," the periods of life thought to be most impressionable (Reamon-Buettner, Mutschler, and Borlak 2008). As social scientists have pointed out, the environmental exposures studied through environmental epigenetic research can be many different things, from food to traumatic events, socioeconomic status to various toxins (Müller et al. 2017). In environmental epigenetics, even habitual behaviors come to be understood as environments that may epigenetically influence the health of offspring (Guthman and Mansfield 2013; Landecker 2011; Landecker and Panofsky 2013).

Environmental epigenetic research has been criticized for its tendency to reduce complex activities into oversimplified characterizations of environments and environmental factors (Darling et al. 2016). But one unintended consequence of an environmental epigenetic approach that might be understood as not (yet) reductive is that laboratories themselves have increasingly come to be understood as containing multiple potentially influential environmental factors. As shown in the animal studies performed by the DeTox Lab, epigenetic research practices aid toxicologists' further attunement to the material, chemical, and affectual aspects of their research settings, which are now understood more than ever as potentially having epigenetic effects. Toxicologists demarcate, isolate, and measure the influence of certain environmental factors while simultaneously perceiving and selectively excluding multiple exposures and multiscalar environments in which animals live, eat, breathe, interact, and reproduce. Environmental epigenetic research both reduces and proliferates environments in and through the laboratory environment.

Laboratory Imaginaries

Laboratories have long been imagined as a place in which scientists attempt to model, replicate, and test ideas about the external world (Latour and Woolgar 1986). As anthropologist Karin Knorr-Cetina (1999) notes, much of laboratory practice is about the interior replication of an exterior condition. However, as decades of social studies of sciences have shown, regardless of this perception, laboratories are places that both model external life and are imbued with it. The laboratory is not an island; it is situated within stratified worlds and hierarchies of knowledge that extend to and through laboratory life (Haraway 1988; Strathern 2005; Traweek 1992).

Toxicology, especially laboratory-based experimental toxicology, is based on a similar premise of replicating conditions and effects in model animal and simulated circumstances. It has also historically been perceived as sandwiched between ethical commitments to protect public health and to preserve corporate-industrial interests. As explored in the work of historian Christopher C. Sellers (1997), the role of the laboratory in the development of environmental health science in the United States through the field of industrial hygiene, a precursor of toxicology, was tied to the growth of the industrial workforce. Industrial hygiene quickly gained a reputation in the early twentieth century for being biased toward workers. Corporations even accused researchers of being anti-industry if they claimed to perceive negative health effects of occupational environments (Sellers 1997).

In such a climate of perceived bias from those in the field observing and gathering data from workers and in factories, laboratory research was proposed as a path toward the depoliticization of industrial hygiene. Moving from the workplace to the laboratory, as well as from human to animal research subjects, a select group of industrial hygienists at elite US universities attempted to make their area of expertise more "scientific." Laboratory-based industrial hygienists challenged the observational methods used to study the negative health influences of occupational environments and relied on techniques of chemistry and physiology to distill the effects of specific toxins suspected of harm. In Sellers's words, "The quantitative chemicophysical terms and techniques of the researchers introduced a new potential for abstraction and generality into the study of workplace causes of disease" (1997, 165).

With these quantitative methods in place, industrial-hygiene laboratories would become arbiters of disputes about workplace safety instead of advocates for a reduction in workplace harm through the creation of (what was perceived as) a neutral zone of knowledge making. As industrial hygienists began investigating toxins through experimental science, they attempted to create environments in and through the laboratory that—unlike factory settings—could, in theory, be precisely controlled. This was done in part as a means of formalizing industrial hygiene as a science, improving the empirical veracity of its claims by isolating causal factors outside the complexity of the "*milieu extérieu*" (Sellers 1997, 165). At this time, nonhuman animal studies became crucial to this physical delocalization from the workplace and the accompanying ethical detachment from the worker (Sellers 1997, 169–70).

Furthermore, as historian Michelle Murphy describes, nonhuman animal experiments made it possible for laboratory-based researchers to perceive what was imperceptible to field researchers: "the specific chain of effects that a chemical consistently caused in human physiology. That is, by passing the connection between chemicals and bodies through the laboratory, industrial hygiene promised to objectify the cause and result of industrial disease" (2006, 87). Many of the techniques developed in these early laboratory-based industrial hygiene and toxicology remain central to toxicology today. The chemicophysical techniques outlined by Sellers and Murphy, such as measuring small amounts of substances in blood and using animals to locate the internal mechanisms responsible for certain adverse effects, continue to be important to toxicologists. The dose-response curve continues to inform policy to varying degrees, depending on the governing body and how exposures and harmful effects are imagined in broader political and social worlds.

But research techniques have also changed. As anthropologists Kim Fortun and Mike Fortun (2005) show, there is a growing sense among toxicologists that the techniques they use must further change to develop more-complex and more-robust renderings of exposures, mechanisms, and effects. Fortun and Fortun describe this as a shift in both toxicology's scientific imaginary and technological means. In their words, "If toxicology in the past could be described as always working in tandem, if not collusion, with industry, today it must be described as working in tandem with industry, with informationally well-supported critiques of industry, with the promise of both genomics and informatics, and with a culture thoroughly caught by 'complexity'" (2005, 50). This complexity capture can be seen clearly in the growth of toxicogenomics (Fortun 2011) and in its aftermath: environmental epigenetics.

In environmental epigenetic studies that rely on animal models, the recognition that multiple environmental factors within the laboratory may influence research results has shifted what Murphy calls a "regime of perceptibility." Regimes of perceptibility, in Murphy's words, "establish what phenomena become perceptible, and thus what phenomena come into being for us, giving objects boundaries and imbuing them with qualities" (2006, 24). Over the last twenty years, toxicology's regime of perceptibility has shifted. As more and more of the material, chemical, and affectual dimensions of laboratory life are perceived as having potentially experiment-interfering effects, such interference becomes more than background noise (Shapiro 2015). Interfering factors become environmental factors, perceived as shaping and changing the bodies, dispositions, and epigenomes of the animals that remain at the center of so much toxicological research. The low-level exposures, multiple exposures, and intersubjective reactions that Murphy points to as imperceptible in toxicology's twentieth century are now often taken into consideration by toxicologists, even if the tools for regulating such complexities are works in progress (Boudia and Jas 2014).

Modeling Exposure

One day in April I came to the DeTox Lab to assist Wang Bo in responding to editorial feedback on an English-language publication he had submitted to a peer-reviewed journal based in the United States. As first and corresponding author, he had just received a "revise and resubmit" and was excited to have another manuscript one step closer to publication. The article was based on an animal study conducted in the lab that used Sprague-Dawley rats to test the influence of exposure to bisphenol-A (BPA) on testicular function. BPA is an industrial chemical mainly used in plastics that has long been known to have "estrogenic effects" and is classified as an EDC (Vogel 2013). Like many EDCs, BPA is thought to negatively influence men's reproductive health in particular, including male fertility (Cariati et al. 2019). At this stage, the DeTox Lab's research was focused on metabolomic effects of exposure on rats orally administered BPA. In the future this study would be extended to include an intergenerational epigenetic component. How did exposure to BPA or other specific EDCs affect the epigenome of future generations? Did exposures to EDCs in one generation have effects on the reproductive health of male offspring one, two, even three generations later?

Like many agencies involved in toxicology, the DeTox Lab often conducted animal experiments, using specific nonhuman species (primarily rats, mice, and zebrafish) to measure the effects of specific toxins. From its early days, toxicology has relied on animals as models to study the potentially harmful effects of the places where humans work and live, as well as the mechanisms of those effects (Sellers 1997). Alternative methods of toxicological research have been developed, both to avert the use of animals and to better capture what is viewed as the growing complexity of data needed by and produced through toxicology (Fortun and Fortun 2005). But even with the growing popularity of these in vitro (cellular) and in silico (digital) methodologies, animal studies continue to be required by many regulatory bodies before human use of specific drugs is allowed or new chemical regulations are passed.

The typical justification for this requirement is that humans are "complicated organisms," so the models used to show effects of toxins must also be complicated organisms. According to professional associations such as the US Society of Toxicology (2006), only through the use of "whole animals" and "living systems" can toxicologists come close to approximating the influences of toxins on humans. Although this argument relies on the idea that nonhuman animals are enough like humans to be used as models, it also implicitly suggests that they are enough unlike humans to warrant their exposure and "sacrifice" or killing (Sharp 2018). In environmental epigenetic research, animal models are also valued as living systems because they dynamically interact with environments "like humans." Although it is true that in vitro cellular models interact with culture medium and other laboratory surroundings and that in silico models could be interpreted as influenced by digital design, animal models are considered a more dynamic and realistic rendering of organism-environment interaction (Phillips and Roth 2019). The use of animals in research is thus justified through the idea of them being more like humans than other research objects but less like humans than other research subjects.

From a toxicological perspective, outside the laboratory humans and nonhuman animals are constantly interacting with numerous environments and taking in a variety of exposures. Such multiple exposures from various sources pose significant research challenges for a science rooted in a single variable dose-response method of research (Murphy 2006). In nonhuman animal studies, however, there is an idea that animal studies offer a feasible alternative to human research (such as epidemiological approaches) because animals are both dynamic and, to an extent, controllable. In the laboratory environment, animals can be bred to certain genetic specifications, be cared for in certain ways, and be exposed to specific chemicals or behaviors but shielded from others. At least, this is the idea.

More recently, however, the laboratory environment is increasingly being understood as a place that itself needs to be managed, if not controlled. Much more than a sterile, neutral, and objective research setting, the laboratory is increasingly viewed by scientists as a space beyond control. As historian Nicole Nelson writes, "The creation of a truly controlled laboratory environment, where all of the parameters that might impact behavior were known and accounted for," has become out of reach (Nelson 2018, 8). The stakes of such awareness are raised when the influence of "environmental factors" on animal models is not only confounding to experiments but is at the heart of research. Increasing research on how environments influence research subjects within and across generations changes the importance placed on potential environmental factors coming from within the laboratory itself. In environmental epigenetics the results of multigenerational experiments involving nonhuman animals have come to be perceived as potentially influenced by a number of laboratory conditions and researcher behaviors: climate, diet, housing, frequency of access to water and food, stress levels brought about by proximity to other animal subjects, and human interaction are often mentioned as environments that might influence an experiment's results.

For example, in the article I was helping to copyedit that day in the DeTox Lab, the details of a number of environmental factors were included. All rats were given a week in the lab to acclimate. Humidity levels were monitored in the room where the rats were kept. Unlimited access to food and water was provided. A number of caveats like these were listed in the article to account for the laboratory environment—suggesting to journal editors, peer reviewers, and eventually a broader audience that researchers from the DeTox Lab were doing everything possible to limit the influence of all environmental factors except one: BPA exposure. BPA exposure was the single variable being investigated and the one that they hoped to test for its metabolic influence on rats, as compared to the control group, which was not administered BPA. But to isolate BPA, many other kinds of environmental control had to be attempted.

In other DeTox Lab publications, methods of control were different yet similarly specific. Articles listed the exact concoction of the culture medium in which epididymal cells are placed; the amount of bovine serum that the albumin culture contained; how long the cells sat in this culture and at which temperature; the exact nature of the substance in which testicular cells sat; how, when, and by whom blood samples were collected, clotted, centrifuged, stored, assayed; and what equipment was used, procured from which manufacturer, and based in which country. These are the details of scientific practices, listed to ensure replicability and rigor. But these methodological details are also increasingly perceived as "environmental conditions." Behaviors, materials, and chemicals used in the laboratory are seen as potential environmental factors that may influence the cells and substances being tested.

Sociologist Carrie Friese and anthropologist Joanna Lorimer describe such "exacting control of the environment (supplies of food, temperature, etc.)" as a kind of care work (Friese and Latimer 2019, 128). They stress that environmental control is a practice that requires attention to detail and even a kind of empathetic relationality between animal and scientist, who performs such care. Care also involves, in the words of anthropologist Lesley Sharp, "sophisticated knowledge of welfare mandates, ever-evolving 'best practices,' and species-specific expertise. For animal technicians, laboratory labor is a moral project" (Sharp 2018, 157). By reframing such environmental control as care work or a moral project, science-studies scholars show that the environments of animals in laboratories are intrinsically tied up in human occupational environments. Further, they suggest that questions about who does care work and how are also matters of institutional and academic hierarchies, including ethnic and gender disparities in higher education more generally (see also Lappé 2018).

Contemporary toxicology research that uses environmental epigenetic approaches both builds upon and diverges from a history of "control" and care through an awareness of the potential influence of the laboratory environment as itself containing multiple environments. As the epigenetic effects of more and more chemicals, materials, and behaviors become perceivable, potential influential environments come into being. Demands for care work are exacerbated not only by the allure of control but also by a sensibility that accepts the impossibility of containing the proliferation of environmental factors despite scientists' best efforts.

Inheriting Effects

On a Wednesday in April, Wang invites me by text to meet him and his graduate student colleagues in the main laboratory across from their shared office. The laboratory room usually smelled of dust and chemical solutions, but when I enter the room today, there is a musty odor. I have to quickly take a tissue from my pocket, covering my nose before I sneeze. As I walk farther into the room, I join the graduate students—many of whom I now recognize—in the northwest corner. I keep the tissue at my nose to cover the smell I realize is coming from stacked cages in the corner of the lab. In each cage is a mouse—small and frantic, suckling from a water bottle and scurrying around its small enclosure, which is densely packed with shavings. Wang Bo reaches into a cage and lifts out a mouse. For weeks he and his colleagues had been waiting for this critter, a third-generation male mouse bred to test for the epigenetic mechanism involved in the intergenerational inheritance of semen decline.

Six other graduate students, primarily women pursuing their master's degrees in toxicology, have gathered in the lab to assist with the experiment.[1] They watch Wang as he takes the mouse from the cage and holds it in his hands. As he walks to the laboratory bench, he explains the experiment to me. This mouse's grandmother, so to speak, is the study's first-generation mouse (F0 in figure 5.1) and was bred with an RNA micro-deletion associated with sperm decline in situations of human direct exposure.[2] F0 was then bred with a "wild-type" male (meaning a male that was not intentionally genetically modified or exposed to toxins). She had a litter, and from this litter, a male mouse (F1) was selected and bred with a wild-type female mouse, who also had a litter. From this litter another male mouse (F2) was selected, and he now sits in Wang's hands. Would this particular mechanism, an RNA microdeletion, result in the anticipated effect—intergenerationally inherited sperm decline in F2? This was the question, and the hope, at the center of today's experiment.

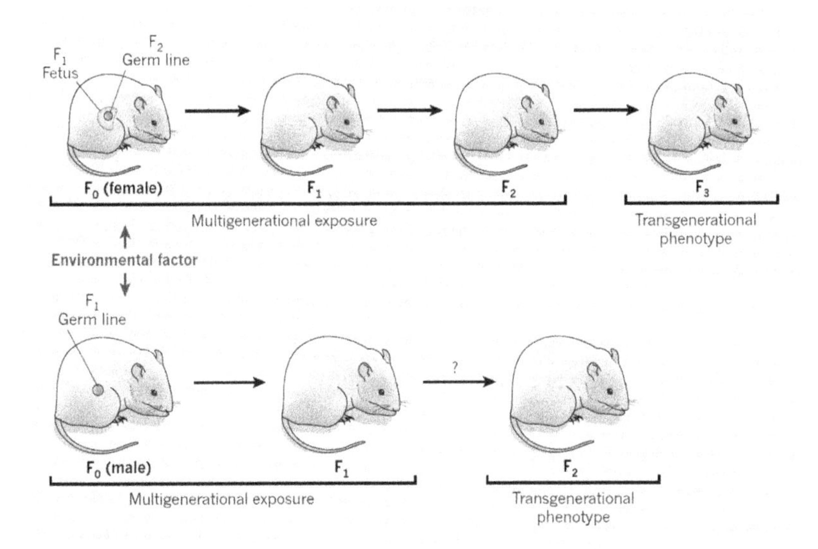

FIGURE 5.1 This image, from an article published in *Nature* titled "Fathers' Nutritional Legacy" (Skinner 2010), was shown during a presentation by DeTox Lab PhD student Wang Bo at a Toxicology Department colloquium. It was meant to help explain his epigenetic research to the crowd of faculty, postdoctoral researchers, and PhD students.

The group now stands in a semicircle facing the mouse on the bench, ready to assist Wang as needed. I take a seat on a stool next to Wang, who carefully holds the mouse with two hands. My eyes and body are squirming, for I am not practiced in the detachment involved in such work (see Sharp 2018), and faster than I can see, Wang has killed the mouse and now dangles its open mouth over a small vial, which quickly fills with blood. "You killed it already?" I ask. "Yes," he answers. "How?" I inquire. "His neck," he responds. I look closer, becoming as uncomfortable as Wang is careful. He shakes the last trickle of blood from the mouth as the tiny creature spasms in midair. I let out a small shriek. "It is dead," he offers to calm me. I nod, embarrassed at my outcry, as he sets the mouse on the table and hands the vial of blood to one of the graduate students who stepped out of the semicircle, each previously delegated a task in the dissection process.

The mouse is positioned on its back, and the skin on its belly is cut vertically and horizontally. Fur and a surface layer of skin are carefully removed before its organs are separated out and placed into separate containers. Wang consults the instructions: he is to remove certain organs and analyze them. The kidney. The heart. One by one he removes body parts, and one by one a graduate student comes to collect them in test tubes clearly labeled and ready for storage until after the primary goal of today's experiment, semen analysis, is achieved. Here, toxicology's living-systems approach makes its way into scientific method. Each part of the body may tell a partial story about how animals interact with environments and exposures. However, Wang is most concerned about isolating the testes and epididymis, a small tube that runs from the testes through the penis along which sperm travels from the inside out.

After being carefully located and removed, the epididymis is placed in a vial like the other organs, but it is immediately brought around the room to the other side of the bench, where a sterilized pair of miniature scissors and tweezers, a petri dish, and another vial are kept. I follow Wang as we move to the next station. Having isolated the epididymis, Wang begins to cut it into tiny pieces with the miniature scissors so the sperm can go out, he tells me. As I stay close to Wang, others work around us. Some clean up the dissection area. Others begin preparing the samples they have been placed in charge of for analysis. However, Wang stays focused on the sperm. Even though he cannot yet see the sperm, he knows they are there and that his ability to perceive them through the computer-assisted semen analysis (CASA) machine relies on his precision. Earlier this morning, results from the CASA had left Wang and colleagues without anything to work with. The epididymis had not been cut finely enough to eliminate tiny chunks of flesh from disturbing the clarity and accuracy of the sperm count. This time, Wang cannot risk delegating this role. He works with patience and exactitude, ensuring that he has cut up the epididymis multiple times into tiny parts.

Immediately after the dissection of the epididymis, Wang asks me to follow him as he makes his way down the hall of the second floor to the stairs, test tube in his hand and close to his chest. He walks briskly as we make our way down to exit the building, explaining as he walks that he hopes to keep the sample warm until we reach the CASA machine, housed in the reproductive medicine building across the street. As we walk, I am struck by a familiarity. While I was working in an IVF clinic before going back to graduate school, some of the men whose semen analyses I scheduled were unable to produce semen samples in the clinic's collection rooms, for reasons that varied from religious to psychological to sexual. The clinic discouraged at-home sample collection but accommodated these men regardless and had an extensive protocol and a closet of special supplies—antibacterial soap, sterile containers, and/or semen-collection condoms, for instance—that I would give to men in small white paper bags to ensure privacy. Part of the instructions I gave were to keep samples as close to body temperature as possible. Many arrived in the Upper East Side by taxi so that they could come straight from the street below up to the clinic without risking the loss of their sample. Men would then sit in the clinic's lobby and wait for a lab technician to call their name and take them to a private space where they could hand over their sample. I would often see men, waiting in the lobby—sample in hand, hand on their chest, keeping their samples warm.

Wang moved from the DeTox Lab to the building across the street with this same nervous and protective energy. For him, the stakes of this experiment were particularly high because of the intergenerational aspect. If something Wang did knowingly contributed to the inability to complete the experiment, like the imperfectly dissected epididymis or the temperature outside, then the breeding process would need to start all over again. It would be more weeks, and more funds, to order the specifically bred mouse, which would then be made to reproduce in house and then be dissected in order to conduct this semen analysis. Wang was aware of multiple environmental factors such as temperature and their potential influences on sperm.

Of course, there were other factors that remained less perceptible to Wang, those that exist "beyond the screen," as Martine Lappé (2018) describes in her study of environmental epigenetic animal experiments. In the DeTox Lab, such imperceptible factors might have been the high levels of particulate matter in the air outside, blowing through the crack in the laboratory window that wouldn't quite latch, located just above the mice. Perhaps the noise and reverberations from the ever-going construction occurring on campus and across downtown Nanjing that sometimes shakes the building heightened stress levels in the mice. Perhaps the fumes from other solvents and chemicals being used in other experiments happening in the same laboratory, while the mice or rats acclimated for one week, altered experimental results. Even when toxicologists ignored or explained away these other components of the laboratory environment in order to focus in on other environmental factors and effects, such multiplicity remains, both perceived and imperceptible.

Proliferating Environments

Wang continues to hold the mouse's semen sample close to his chest as we wait in the lobby of the reproductive medicine building, a five-story complex across the street from the DeTox Lab. Large glass windows provide light to the interior foyer and hallways that form a rectangle around the atrium upstairs. The contrast between the toxicology building we just left and the building we have entered is striking—the inequality between the financial benefits of treating infertility and researching infertility's cause manifest in architecture and equipment. We are here because Wang has arranged to use the department's CASA machine. A graduate student in the reproductive medicine program has been put in charge of assisting us while we use the machine, and we meet him in the downstairs lobby.

Together we take the elevator to the fourth floor, then walk around the atrium to another laboratory. On top of the machine sits a tiny wooden box with a metal hook latch. The graduate student carefully unlatches the case with hands covered in rubber gloves, revealing a stack of platelet slides. Wang puts on his own rubber gloves and retrieves a pipette full of the mouse semen out of the test tube and puts it on a platelet slide. The other graduate student covers the slide with a plastic rectangular slip and inserts it into a tray on the machine. Within minutes an image pops up on the machine's small screen. We see sperm sputtering and moving about, another familiar image—perhaps from my days at the fertility clinic, perhaps from the frequent circulation of such reproductive imagery in popular science and health education (Martin 1991; Morgan 2009).

Pressing a button on the control keypad, the graduate student assisting us (who was the only one allowed to touch the machine) moves between five captured images. He chooses one with Wang's help, and the machine then analyzes this image. All of the measures that laboratory technicians conduct under microscopes—quantity, mobility, motility, viscosity—are here calculated by the machine in less than ten seconds. The numerical values demarcating various aspects of semen quality and quantity then pop up on the screen. Wang takes a digital camera out of his pocket and takes a photo of the analysis results. USB drives are not allowed for fear that the machine may contract a virus, he explains. The activity is repeated four times, with four parts of the sample. One is still "too dirty" (*tai zang le*)—meaning that there are too many particles of flesh mixed among the sperm—but only three are needed to meet the specifications of the experiment.

Having completed the semen analyses, the remaining semen sample is discarded, and we walk back to the toxicology building. What was once a precious fluid (Moore 2007), holding the potential traces of epigenetic inheritance, has now become biohazardous waste. Its digital capture is enough to secure its place as data in this experimental work. These images will now be uploaded to Wang's work computer, their values input into spreadsheets. Wang will spend the next few days working with his colleagues to analyze the blood and the other organs that he removed from the mice. Then statistical analyses will be run to see if the experiment isolated the correct "mechanism of the effect," the mode of epigenetic transmission.

At the end of the week, junior members of the DeTox Lab involved in the experiment celebrated their hard work with a group outing. We gathered for karaoke and then a long dinner in a favorite restaurant as a means of celebrating, regardless of the experiment's outcome. Whether or not the results of the study would turn out to be statistically significant (which we later found out were not), there were a number of steps involved in getting from here to there: study design, ordering specific genetically modified mice and other necessary supplies, waiting for the mice to have offspring as they are kept in the animal breeding department, and then finally acclimating them to the laboratory space, controlling the temperature and humidity of the room, making sure they have food and drink to the specifications, and limiting other factors that may affect them—light, sound, activity. Then all the steps of the dissection process from the time Wang lifted the male mouse from the cage with his two fingers. All these steps, all this time, and all this environmental control now eventually come down to a set of numbers.

An argument could be made here about how a complex situation and process comes to be represented by such singular statistics, about all the things that will be written out of this experiment on spreadsheets and in publications (Latour 1987). This is an argument of reductionism—of how the complexity that environmental epigenetics is meant to capture becomes black-boxed or oversimplified. But there is another argument to be made as well about how such experiments not only reduce such complexity into singularity but also proliferate environments along the way.

In an effort to rethink reductionism, anthropologist Cori Hayden has written about the propensity of social scientists to point out how science reduces complex and situated knowledge into isolated matters of fact: "What if the particular operations that we take as so characteristic of both science and capitalism—reduction, isolation, atomization, abstraction, substitution, interchangeability—are not reducible to themselves? What might it do to the tenor of our arguments to get inside such operations with a certain amount of ethnographic and analytic openness, perhaps finding proliferating relations of same/not the same instead?" (Hayden 2012, 281).

By paying attention to the practices involved in the DeTox Lab's epigenetic experiments, one can see that the lab members both reduce the environment to factors that they attempt to control *and* multiply environments by understanding many of the materials, chemicals, and conditions of animals' daily lives, in and outside of the laboratory, as environments that might have effects in and across generations. Environments, typically one at a time, take center stage in particular studies—this is the isolating part. But if one looks closely, we see that isolation is just part of what is happening. Other environmental factors are recognized in tandem with the isolated factor through practice. Some of this recognition ends up in the literary representation of experiments—humidity levels are recorded, food and water access described. Some occur only in laboratory practices—sperm is kept close to one's chest as you walk quickly across the campus road.

Returning to Knorr-Cetina's interpretation of laboratory practices as ideally the interior replication of exterior phenomena, such replication or modeling is often interpreted as mimicry, a copy not as robust or complex as that which occurs externally or "in real life" or "in humans." Animal models and laboratories themselves are, in Hayden's language, "simultaneously the same, and not the same" (2012, 271). Such internal replication could be interpreted as a kind of insufficient translation or reductionism. But it could also be interpreted as a form of proliferation.

A lot of work goes into demarcating and stabilizing "the environment" in environmental epigenetic research. In such practices you can find the cordoning off of complexity, but you can also find the proliferation of contextual, chemical, and material worlds. In such proliferating practices, environments are not simply recognized, where once they were ignored through a reductive scientific lens. Environments actually come into being, in multiple, overlapping, partial forms (Mol 2002; Strathern 2004).

Perceiving and Managing Complexity

Research in the biological sciences is increasingly defined by postgenomic forms of complexity (Gibbon et al. 2018). Environmental epigenetics draws attention to the way that environments—variably and multiply defined and demarcated—influence epigenomes in present and future generations. As more and more potentially intergenerational effects of more and more environmental factors are perceived, how do you isolate casual associations? The complexity of environmental exposures is not just a theoretical hypothesis but a methodological concern as well. Such a moment invites social and natural scientists to become more openly reflexive, or at least more forthright, about the work involved in the momentary capture of laboratory environments—including the physical, chemical, and interpersonal dynamics within these spaces.

Through attuning to the detailed procedures that make and unmake environments of exposure in and outside the laboratory environment, one can see how toxicologists ascertain and manage the complexity they have become more open to perceiving. The environmental management that occurs during the design and execution of laboratory experiments, particularly in environmental epigenetic studies, brings about an intensification of attention to certain aspects of what constitutes the laboratory environment and all that must be done to achieve the ideal of replicability through the scientific method. The critique of the (toxicology) laboratory as a space of simplification, isolation, and reduction of complicated information is itself then complicated by the increased perception of complexity in epigenetic research (Nelson 2018). As part of a growing trend toward epigenetic research approaches, such heightened attunement has resulted in the proliferation of environments, in and outside of the lab.

Coda

Since the time of my research, China has developed a new double-edged reputation—as both a reckless polluter negatively influencing international environmental health and a global leader in sustainable development and climate-change action. Such a contradiction plays out in specific ways. For instance, in 2018 the Chinese government announced that it would aim to achieve zero growth in pesticide use by 2020. As this deadline approached, several pesticide factories that did not meet environmental standards were shut down, a potential achievement for environmental activists and everyday citizens. But the story does not end there.

In Changzhou, the same city where the DeTox Lab initially studied the genotoxicity of pesticides to human sperm, a school was built near the former grounds of multiple pesticide factories. Shortly after its opening, many students began to experience strange symptoms: rashes, coughs, headaches. Biomedical diagnoses of students' conditions in subsequent months ranged from the minor—eczema and dermatitis—to the more severe—lymphoma and leukemia. In these conditions, Changzhou's industrial past was living on in the present, the aftermath of chemical production given new life in children's bodies (Makinen 2016).

Such examples of toxic inheritance have led toxicologists in China and elsewhere to explore questions about the toxic afterlives of environmental exposures through epigenetic research in laboratory and digital settings. They are also increasingly conducting birth-cohort studies, which take a more epidemiological approach to questions of disease inheritance. Birth-cohort studies are characterized as longitudinal investigations of research subjects with at least one common characteristic: usually being born in the same time and place. Such studies are increasingly common around the world and across numerous disciplines (Gibbon and Pentecost 2019). The DeTox Lab was just beginning to conduct cohort studies when I was there in 2011. At the time the potential of this research methodology loomed large in the scientific imaginary of the lab, promising a means of including more people from multiple generations in research on an increasing number of health conditions, resulting from the many toxic exposures faced by those living in and around the Yangtze River Delta. Today, as I analyze from afar the research that the DeTox Lab has undertaken since I returned from my fieldwork, it is striking to think about how the methodology of cohort studies has transformed the lab's research.

With the further addition of birth-cohort studies to the DeTox Lab's scientific repertoire, new questions have become possible, as has a new scale of research. For the lab, cohort studies were understood as a methodology of casting the widest net and of quickly asking many open-ended questions about correlations and associations between reproductive and developmental conditions and the many indirect and multifactorial exposures faced by the lab's research subjects. This is seen as urgent work in a context where people often face levels of everyday exposure to toxins that far exceed many other national contexts, as was shown to be the case in the lab's earlier sperm research that found levels of exposure to specific EDCs in China to exceed those of other nations by as many as eight times. It is also work that is now broadly supported through national and international funding mechanisms that are part of a globally relentless focus on prenatal, fetal, and infant health, especially in the first thousand days of life (Pentecost and Ross 2019).

Whereas the DeTox Lab's research on male infertility primarily stressed how sperm health was influenced by occupational, national, hormonal, and dietary environments, birth-cohort studies often fold multiple external environments into the singular maternal transmitter and fold mothers into their offspring (Mansfield 2017). This scalar reductionism simultaneously overlooks maternal health *and* places the burden for the effects of the "maternal environment" on women (Landecker 2014). Cohort studies' concentration on the mother-infant dyad problematically reasserts an idea of an individualized, maternal responsibility for what might be more accurately described as the effects of capitalist industrialism. In birth-cohort studies, the cause of China's "fetal abnormalities" reads more like a failure of mothers to protect and reproduce healthy children than the unregulated industrial development and economic growth articulated in the DeTox Lab's earlier research.

This is not only the case in the DeTox Lab's more recent research; it is also increasingly true of environmental health research more generally. As the stakes of environmental damage increasingly become perceived as a matter of securing livable futures for subsequent generations (Dow 2016; Cohen 2020), environmental health sciences are increasingly thinking of environmental harm as inheritable. Multidisciplinary and interdisciplinary investigations of how chemical toxins affect biological systems are increasingly being conducted through approaches such as environmental epigenetics, developmental origins of health and disease (dohad), and children's health. Each of these research frameworks highlights how environments affect intergenerational health, using the word *environment* to describe a variety of exposures and circumstances. In such research, connections between people and between people and "environments" are often considered the source of disorder, dysfunction, and disease: intergenerational relations are framed as a potential conduit for the damaging effects of environmental exposures (E. Roberts 2017).

In toxicology in particular, over the last decade researchers have been devoting an increasing amount of time and money to studying the intergenerational dimensions of toxic exposures. The subfield of "reproductive and development toxicology," once the home for those interested in the effects of toxins on reproductive and developmental health, has overflowed into much of the discipline. This has given way to what is now being called "generational toxicology," meaning the study of "the impacts of environmental exposures on subsequent generations" (Kubsad et al. 2019, 8). A trend toward generational toxicology (under this title or not) is happening in studies around the world, resulting in the formation and multiplication of multigenerational animal experiments and birth cohorts, as well as the development of informational techniques to analyze the potential of intergenerational transmission of toxic exposures through data modeling.

As toxicologists increasingly turn to birth-cohort studies to conduct investigations on the potential intergenerational harms of toxins, researchers capture and produce anxieties about the un/making of future kin in a world increasingly saturated with toxins. Such toxicological accounts of intergenerational harm work through overly simplistic depictions of reproduction and biological relatedness, often depicting intergenerational connections as the heteronormative blood relations long featured in anthropological accounts of kinship (Franklin 2013b; D. Schneider 1968). In its further embrace of the intergenerational in human birth-cohort studies, the field of toxicology more generally seems to increasingly imply that the most crucial component to health is "normal" reproductive "success." As Anne Pollock argues of endocrine-disruption discourse, "The emphasis on the intergenerational and on genetic continuity is an undue limitation on queer possibility" (2016, 187). It is also, I would argue, an undue limitation on kinship imaginaries more generally.

In toxicology the oversimplification of the ways in which one generation begets another and the valorization of biogenetic relations and bodily reproduction make clear that the intergenerational lens has the potential to take on gendered and heteronormative assumptions. Ideas about who or what constitutes and interconnects a generation, as well as the boundaries of reproduction (Murphy 2013) and the standards of "normal" development, are fundamental to toxicology's notions of what constitutes toxicity and harm (Di Chiro 2010). And so is a kind of "reproductive futurism" that valorizes the child and biogenetically related kin as the source of meaningful continuance (Davis 2015; Edelman 2004). The narrow approach to the intergenerational practiced by toxicology is but one more reason for social scientists to approach the field with ambivalence. As Alex Nading (2020) notes, ambivalence toward toxicology based on its normalizing tendencies has resulted in anthropologists taking both friendly and adversarial stances toward the field. These normalizing tendencies include the question of how toxicologists approach who counts as kin.

For more than a hundred years, anthropologists have documented and analyzed patterns and practices of human relationality around the world, purporting theories of kinship as a universal means of social organization and arbiter of social cohesion through exchange. Even after the so-called death of kinship, when David Schneider (1984) and feminist anthropologists (Collier and Yanagisako 1989) expressed skepticism about the ability of kinship studies to capture "actually-existing biological relatedness," anthropological kinship studies continued, moving on to, as Sarah Franklin describes, "the wider question of how kinship models enact or perform culturally specific ways of knowing about the world at large" (2013b, 286). Still, for many the relevance of "kinship" seems to slip out from the center of anthropological analyses, becoming "abstract and dry" (Carsten 2016) or in need of constant redefinition (Sahlins 2013). But while anthropology continues to wrestle with how to relate to the history and contemporary life of kinship studies, scholars in adjacent fields are turning to kinship as a lens through which relationality might be reimagined in order to secure more-livable futures.

At a moment of grave concern for the future of our planet, social scientific effort has been made to study and promote understandings of nonhuman beings and entities specifically as kin. This is certainly part of a broader effort to think beyond the human when discussing "alternative kinship imaginaries" (Ginsburg and Rapp 2019) in queer, disability, Black, and Indigenous studies (Benjamin 2018; Kafer 2019; Luciano and Chen 2015; TallBear 2018). But the recent promotion of kin making more specifically focuses on seeking out alternative kin at a moment of planetary urgency. For example, Adele Clarke and Donna J. Haraway's (2018) edited volume *Make Kin Not Population* suggest kin making as a strategy for multispecies co-flourishing and increased repro-environmental justice, while Deborah Bird Rose (2017) describes nonhuman and "other-wordly" kinship as a means of connecting across temporal, spatial, and taxonomical divides at a time of extinction. Making kin has also been presented as a hopeful pursuit at a moment of ruination (Tsing 2015), as the paradoxical and messy entanglements between humans and nonhumans (TallBear 2015; Todd 2017), and as the lens through which harm to Black communities might be understood and kinship in the aftermath of violence might be cultivated (Benjamin 2018).

Such understandings of kinship move beyond its ancestral focus (Balayannis and Garnett 2020), stretching anthropological uses of the concept and pushing anthropological studies of the environment in new directions such as chemical kinship. Anthropologist Vanessa Agard-Jones (2016) describes "chemical kinship" as a connectivity based in shared exposure through which communities of injury find the grounds to demand accountability (see also Petryna 2002). Such ambivalent renderings of kinship also encourage us to reimagine our potential relationships to toxicity or "the longstanding relationships and emergent social forms that arise from chemical exposures and dependencies" that Eben Kirksey and Nicholas Shapiro call "chemosociality" (2017, 484). How might such frameworks bring into view intergenerational relations beyond those most often conceived in and through toxicology? At a time of heightened attention to the environment from a variety of actors, might a focus on intergenerational health also have the potential to catalyze creative understandings of toxic relationalities and responsibilities? The future in/fertility of such environmental work is yet to be known.

Epilogue

After completing my doctoral dissertation based on fieldwork in China, I started a postdoctoral research appointment with the Reproductive Sociology Research Group at the University of Cambridge. While in this position I wrote the book proposal for this manuscript, which described the way that the DeTox Lab's research contextualized and complexified ideas of reproductive and developmental health causality through multiple environments. I also submitted and revised my first academic article—an earlier version of chapter 4—about the DeTox Lab's generous understanding of the maternal environment as a more-than-human, intergenerational space that had the potential to redistribute maternal blame. But as I was drafting and reworking these documents, I was also coming to a new understanding of the stakes of my claims. I was now pregnant, and my body was being surveilled in ways both familiar and strange.

During my pregnancy I experienced the gendered pressure to protect and care for my fetus and prepare for their arrival in an age of "intensive parenting," when child rearing includes an ever-expanding list of responsibilities (Faircloth 2014). I was offered prenatal genetic screening and refused to have the chromosomes of my fetus "read," as Rayna Rapp describes. I thought about her book *Testing Women, Testing the Fetus* as I tried to make sense of my decision as an example of why "a routinizing technology does not always stay en route" (Rapp 1999, 167). I felt the burden of the mental load building as I made plans for midwife appointments and birth classes, household supplies and equipment, and maternity leave. I was given advice by strangers about everyday habits, risks, and time management.

But based on my understanding of the way that mothers in particular have been made to feel responsible for the health of their fetuses and children, there was another component of such pressure that seemed newly amplified: the fear of toxic exposures and their intergenerational effects. Of course, I was noticing the effort to protect developing fetuses and children from toxins with potential reproductive and developmental effects partially because I studied them. But the discourse seemed to be everywhere. Increasingly pregnant women, and even women who are "pre-conception" (Pentecost and Meloni 2020), were being advised to not only eat well and exercise regularly. They were also told to avoid toxins—heavy metals, pesticides, volatile organic compounds (VOCs), bisphenol-A (BPA), cosmetics, gas fumes, and other everyday sources of exposure identified as potential threats to fetal health.[1]

For me, such warnings meant that my already cautious approach to consumerism, which involved buying "organic" and avoiding plastics, was exacerbated by pregnancy. I found myself not only following my midwife's recommendations to cut back on caffeine, sugar, and raw fish but also buying organic baby clothes and mattresses, shopping online for low-VOC baby furniture, and adding green-certified baby toys to my wish lists, as the "natural mama" blogs implored. I was trying to buy my way out of the world my future baby would soon inhabit through a conscientious consumerism that I was aware would do little to change the shape of my little one's environments in the long run. And I was not alone.

Since the time of my fieldwork, multiple anthropologists have documented the ways that an emphasis on the intergenerational effects of toxic exposures in environmental health sciences such as toxicology, as well as public health, has ignited a wave of toxic protectionism among mothers. These forms of protectionism range from the widespread wearing of radiation vests in China (J. Li 2020) to the avoidance of pesticides and nonorganic foods (MacKendrick 2018) to the advocacy efforts of moms who fear the aftereffects of the Fukushima nuclear disaster (Kimura 2016), an event that occurred while I was conducting fieldwork in Nanjing. The efficacy of such protectionism is never straightforward, but what is clear is how logics of maternal responsibility are increasingly enmeshed in questions of environmental toxicity; the onus of safeguarding future generations from environmental threats is unequally distributed along gendered lines. Furthermore, the concern about the effects of toxicity often foregrounds the consumer and her family, disregarding the potential harms faced by those who produce such products and their kin (M. Chen 2012; Nash 2006; Rojas and Litzinger 2016).

As my first accepted article about epigenetics' potential to deindividualize maternal blame entered the final editing stage, my hopeful rendering of the DeTox Lab's epigenetic approach seemed optimistic, if not idealistic. However, even this new interpretation of the article's likely shortcomings, its inability to describe the persistence of rhetoric that blamed the mother and responsibilized the individual in China and elsewhere, took another unexpected turn shortly after I gave birth. I had what is medically referred to as a minimally invasive childbirth, which many mothers or advocates would refer to as "natural birth." My son was born into a pool of water at a birthing center, and I delivered him with the help of my partner, gas and air, hypnotherapy training, and a midwife. I was released from the hospital the next day with some difficultly breast-feeding but otherwise ready to return home. But three days later, in a haze of sleepless nights and failed attempts at feeding, my naturalist ideals were tested when my son jaundiced severely.

During the visiting midwife's day 3 visit to our rental home in the outskirts of Cambridge, she urgently requested we take our baby to the accident and emergency wing of the same hospital where I had given birth. After a fearful drive and an initial assessment, my son was quickly transferred to the Neonatal Intensive Care Unit (NICU), where he was isolated and connected to an IV and a feeding tube. There he was given antibiotics, a spinal tap, and pinprick blood tests on his heels every few hours, and he was placed under special UV lights. After days of hearing the midwives emphasize "biological nurturing," "the fourth trimester," and "creating a nest" where mama and baby remain as closely and "naturally" connected as inside the womb, this was the most artificial environment I could imagine. His body was so fundamentally intertwined with the synthetic that, at first, he could not be held—only comforted with a distant voice and a highly sanitized touch. Reflecting on his own child's neonatal circumstances, science and technology studies scholar Jody A. Roberts writes that "for better or worse, our lives are dominated, and in a tangible and real way made possible, by the very plastic that I've tried so hard to avoid and protect us from in our daily lives" (2010, 112).

My intimate postnatal experience of the paradox of plastics and the synthetic but lifesaving quality of the NICU and its artificial materials was much like Roberts's. Luckily for my partner and me, my child's situation was not as serious as originally thought. His severe case of jaundice was not caused by a blood incompatibility, as doctors originally feared, but was simply a matter of "breast-feeding jaundice," which occurs because the baby has not received enough milk. Unlike the situation faced by Roberts, my baby's dilemma was easily remedied with breast milk and formula and intensive UV lights that helped dissolve the bilirubin that had built up in his system. But the reasons that such jaundice occurred, the reasons that I did not feed my baby enough formula when I couldn't get him my breast milk in the first days of life, again relate back to Roberts's work, specifically to the phobia he describes surrounding plastics and toxic synthetics more generally.

Most cases of breast-feeding jaundice are resolved at home as the baby is fed and placed in sunlight. In our case our baby was not being fed, despite my concerted effort. I had not "properly established breast-feeding" in the hospital before I left; my infant was unable to latch on to what the midwives diagnosed as "flat nipples," a condition commonly estimated to affect about 10 percent of women. Flat or inverted nipples can create breast-feeding difficulty that can often be remedied through several predelivery techniques or after delivery, if the nipple is pulled out with the baby's powerful suction. If neither of these strategies works, a nipple shield—a small device that is placed over the nipple when the baby is feeding—can allow a baby to latch and facilitate milk transfer. However, the use of this small device is often not recommended by midwives, and others such as La Leche League, without professional guidance. And even with professional guidance I was discouraged from its use, and from using formula unless absolutely necessary. Despite my critical thinking around motherhood and unrealistic ideals of "natural" childbirth, this professional advice exacerbated my underlying fear of the synthetic. Formula became the enemy of breast milk, nipple shields the enemy of breast-feeding.

Similar to Roberts', my family's experience in the NICU led to a resignation of an oversimplistic "toxiphobia" and to a more nuanced understanding of the paradox of plastics. But my experience was also filtered through my own positionality as a new mother. Those few but long days watching my child under the UV lights gave me pause. The diagnosis of "breast-feeding jaundice" and the recognition that my own inability to see past the naturalist rhetoric, to accept the paradox of plastics, may have contributed to his condition was overwhelming and new. I was experiencing firsthand what anthropologist Emily Martin (1994) calls "empowered powerlessness." Martin notes that this feeling "of being responsible for everything and powerless at the same time" is partially the product of the body increasingly being understood as a complex system:

> Imagine a person who has learned to feel at least partially responsible for her own health, who feels that personal habits like eating and exercise are things that directly affect her health and are entirely within her control. Now imagine such a person gradually coming to believe that wider and wider circles of her existence—her family relationships, community activities, work situation—are also directly related to personal health. Once the process of linking a complex system to other complex systems begins, there is no reason, logically speaking, to stop. (1994, 122)

Here, as in the epigenetic research that I studied, complexity has the potential to emphasize a person's powerlessness. But, as Martin shows, there is also a possibility that a person might feel that such wider and wider circles of existence are not only affecting her, but also partially within her control. In my case, there was no end, logical or otherwise, to the responsibility I felt for having put my son, my partner, and myself through the NICU experience. I felt responsible not only for growing and birthing a healthy baby as naturally as I could but also for falling into the trap of wanting my baby to have such a natural start that I refused to feed him what I considered to be a synthetic substitute for breast milk or to use a synthetic nipple shield, thereby jeopardizing his life. Complexity had not brought me comfort; it had exacerbated guilt.

I still feel anger about this moment, an anger that again cuts across many levels: at myself for not being able to reflect on the rigidity of my naturalist ideals enough to just give him ample formula; at the midwives who cared for me after birth and came to my home in the days that followed for discouraging me from using a nipple shield when my flat nipples made it difficult for my son to latch; at those who continue to stress that "breast is best" instead of "fed is best" to the extent that formula seems like poison and the window of opportunity to breast-feed so narrow; and at the formula industry for putting profits above health to the extent that there is such a mighty backlash against this sometimes life-sustaining substance.

My fear of formula, fear of toxins, and fear of synthetics, plastic bottles, and nipple shields were supported in many ways. And it was this fear that eventuated in my son's most intimate encounter with both near death and the artificial materials and substances that allowed him to live. Like Roberts, I am now grateful for the plastics, formula, and synthetic lights that pulled my child through. But as someone who is continuously reminded of the intergenerational dimensions of toxic exposures and the threat of toxins to developmental health, such appreciation is fleeting.

Today when, just as the NICU nurse promised, those early days in hospital have become merely a blip, I still struggle to not feel responsible for protecting my child from everyday exposures: flame retardants on his car seat and in the Christmas pajamas he received from relatives, the BPA in the plastic toys he puts in his mouth, the VOCs emanating from our newly painted walls, the pesticides on the fruits and vegetables he eats, and the insecticides and weed killers used on our street, in the parks, at his school. I have read so much toxicology that calls for further research into reproductive and developmental impacts of the various toxins found in everyday items that I see them everywhere. I have trouble accepting a relationship to toxicity that embraces the Pharmakon (Martin 2006; Shapiro and Kirksey 2017) or finds perverse joy in the effects of our human-polluted environment (Pollock 2016). I'm not sure that I will ever be that fearless or that bold.

But at least I can hope for, and join many of my colleagues in calling for, a redistribution of the burden of empowered powerlessness and a reallocation of the pressures that come with the paradox of both plastics and parenting along less gendered, more communal lines.

NOTES

Preface

1 Sarah Franklin (1997) has written extensively about how infertility is often experienced as a disruption to an assumed life trajectory and how this later makes IVF the obvious choice, for this treatment is framed as "a helping hand" on a well-trod linear path.

2 Similarly, in a binational study of women in the United States and Israel, anthropologist Marcia Inhorn and colleagues show that reasons for pursuing "elective egg freezing" do not typically include career decisions, as often assumed. Instead, 85 percent of those who froze eggs stated "lack of partner" as their primary reason for pregnancy delay (Inhorn et al. 2018).

3 The commonly used description of cervical mucus as "hostile" to sperm is yet another example of how the imagery and vocabularies of war make their way into gendered notions of the body (see Martin 1991).

4 Many scholars have written about how biology and biological relatedness are not fixed notions but are reimagined in practice—for instance, through reproductive technologies (see Franklin 2011, 2013a; Hayden 1995; Thompson 2005).

5 Scholars have written on the incongruity of male and female infertility-treatment experiences (Barnes 2014) and the incongruity of male reproductive science and medicine, including gamete donation (see Almeling 2011; Almeling and Waggoner 2013; Martin 1991).

6 Male infertility and particularly a decline in sperm quality have now been linked to age through ideas of paternal effects and epigenetics. For the history of such findings and reflection on why it took so long to seriously consider the role of men's sperm health in infertility and reproductive research, see Rene Almeling, *GUYnecology: The Missing Science of Men's Reproductive Health* (2020).

Introduction

1 See Vincanne Adams, Kathleen Erwin, and Phuoc V. Le, "Governing through Blood," for a more thorough account of the social, cultural, and political factors surrounding donation in China (Adams, Erwin, and Le 2010), and Ruth Rogaski, *Hygienic Modernity*, for a historical account of the importance of semen (*jing*) in late Qing understandings of health (Rogaski 2004).

2 Other important anthropological texts on practice include "Theory in Anthropology since the Sixties" (Ortner 1984) and the follow-up "Theory in Anthropology since Feminist Practice" (Collier and Yanagisako 1989).

3 The term *China* is used throughout this book to reference the People's Republic of China in a way that admittedly elides engagement with the more-complex histories and politics of how the idea of China—as a nation, a geographic region, and an imagined community—is itself shifting and processual. For more on how China itself is a material and symbolic instantiation of broader values, practices, and borders, see, for instance, the work of Michael Kohrman (2005), Xin Liu (2012), Erik Mueggler (2001), Hentyle Yapp (2021), and Wen-hsin Yeh (2008).

4 The term *ecological civilization* was introduced in the late twentieth century, then incorporated into Communist Party policy documents in the early 2000s. In 2012 Hu Jintao reignited the use of the term, incorporating it into the second work report of the 18th Party Congress and then a constitutional amendment. The term has since been put forth as an alternative development strategy that, building on previous civilizing campaigns, takes more than economic growth into account (see Goron 2018; Zee 2020).

5 All personal names and names of research groups, as well as some institutional names, have been changed to pseudonyms to protect the privacy of those I researched. In addition, the details of some experiments and investigations have also been altered or not specified.

6 Even within US and UK biology there are some notable exceptions to this history (see Keller 1984; Franklin 2007).

7 As part of this campaign, the CCP's Propaganda Department and the Chinese Academy of Sciences (CAS) led an investigation into theories of heredity, resulting in the Qingdao Symposium on Genetics. The symposium was a fifteen-day event that brought together more than 130 people, including 48 senior geneticists, agricultural breeding specialists, taxonomists, and embryologists (P. Li 1988; Jiang 2017).

8 The first cytogenetics laboratory in China was established in 1962 at the Institute of Experimental Medicine in Beijing, and in 1963 a division of medical genetics was established at Peking Union Medical College (Luo 1988).

9 Extensive epigenetic research on this famine now studies the increased likelihood for a range of conditions, including high blood pressure, obesity, and schizophrenia, in subsequent generations of those in utero between 1959 and 1963.

10 There is also increased demand for prenatal screening, genetic counseling, and consumer genetic testing in China, although the predictive power of natural talents and abilities remains broadly interpreted as deeply influenced by various environments—familial, educational, occupational, etc. (see Sui and Sleeboom-Faulkner 2010). Inborn (*xiantian*) and acquired (*houtian*) are often regarded as interconnected phenomena (see W. Zhang and Sun 2015).

Chapter 1: The National Environment

1 The shift in causal factors *from* genes *to* the environment that Carlsen and colleagues describe here aged poorly, for gene-environment interaction soon after became a paradigm that scientists from a variety of fields began to embrace (Shostak 2005).

2 The meaning of *positive* here is meant to be juxtaposed with negative eugenics, most infamously practiced in Nazi Germany, where people with characteristics labeled undesirable were removed from the gene pool—killed and/or sterilized. Jiang's characterization of a historical emphasis on positive eugenics in China might be reassessed in light of the more negative eugenics policy that went into place through the mother and infant health program's restrictions on reproductive-technology use, discussed further in chapter 2 and by anthropologist Jianfeng Zhu (2013).

3 Some exemptions to birth-planning limitations did exist, and women resisted and shaped formal policies through informal work-arounds (Greenhalgh 1994). But the "one-child policy," as it came to be known around the world, was enforced primarily via women's bodies through female contraceptive implantation, dangerous and/or repeated abortions, and a massive sterilization campaign. This policy also resulted in a skewed birth ratio, as high as 118 boys to 100 girls in 2011, according to anthropologist Susan Greenhalgh, who argues that its effects were felt across genders. Girl children were missing, the result of infanticide and abandonment. Many unmarried men, especially poor men in rural areas, were unable to marry or have children, so they became "bare sticks" (*guanggun*). Greenhalgh argues that these men were stripped of the conditions "essential for social and even physical survival": getting married and being able to become a "real Chinese man" by fulfilling one's familial duty to have children (2013, 133).

4 For example, state funding for the development of IVF was originally provided through an application titled "Eugenics: The Protection, Preservation and Development of Early Embryos" (Jiang 2015).

5 Like many studies of reproductive technologies, Handwerker's account focuses on female infertility. Although this focus has been justified by many of the factors I discuss in the preface, including women's burden for infertility, the exclusion of men in such work in some ways reifies their absence from infertile blame (Almeling 2020; Barnes 2014). Still, Handwerker shows that the stigma of infertility

clings to women's bodies even if male-factor issues were involved in the cases of many of those she interviewed.

6 China is often painted as "the Wild East" of scientific and medical experimentation—for instance, in the case of stem cell research and more recently CRISPR (clustered regularly interspaced short palindromic repeats) gene editing (Ong and Chen 2010; Song 2017; Thompson 2013; J. Y. Zhang 2012). However, the limitations on who could be treated and through what methods made the use of reproductive technologies in China seem comparatively conservative. At least on paper, government control of reproductive technologies made the Chinese Communist Party's restrictions and monitoring much more extensive than government regulations in other national settings. As someone familiar with how often reproductive technologies were used in the United States to creatively rework normative reproduction, I was also struck by how such restrictions would limit use. For many of the couples I had interacted with at the fertility clinic where I worked in the mid-2000s and for many scholars of reproduction, reproductive technologies were a means of pushing the limits of what is considered biological relatedness through technological means, thereby relativizing and even reclaiming the biological (Franklin 2013a; Hayden 1995; Thompson 2005). In China the use of reproductive technologies, at least by the letter, was closely monitored and regulated alongside pregnancy and childbirth. Such restrictions were designed not only to ensure low population numbers and high population quality but also to ensure "stability" (*wending*) by reproducing a heteronormative family structure (Greenhalgh 2013). These guidelines make explicit what so many feminist critiques of the family reveal: the reproduction of a stable nation relies on the reproduction of a traditional family structure, as well as the gender roles and uncompensated reproductive labor that accompanies this structure.

7 Hypospadias is a condition where the opening of the penis is on the underside of the organ. In later studies, a coauthor of the Carlsen et al. 1992 publication, Nils Skaakebaek, and colleagues also found increased hypospadias to be associated with EDC exposure.

8 Even more recently, public health experts continue to advocate for increased attention to sperm DNA, particularly because of the increasing use of ICSI and the lack of information about sperm DNA's potential intergenerational effects (Perry 2015).

9 Many other scholars have been inspired to extend Lock's local biologies concept in a variety of ways, including stigmatized biologies, situated biologies, subaltern biologies, diverse biologies, global biologies, and collective biologies (Bharadwaj 2013; Fullwiley 2012; Horton and Barker 2010; Hsu 2009; Wanderer 2017; Wentzell 2019).

Chapter 2: The Hormonal Environment

An earlier version of this chapter was published in 2018 as "Swimming in Poison: Reimagining Endocrine Disruption through China's Environmental Hormones" in *Cross-Currents: East Asian History and Culture Review* 8 (1).

1 Like all of my meetings with professionals both within and outside academic settings, this meeting with Greenpeace was partially possible due to my own positionality not only as a white women from the United States but also as a doctoral student with affiliation at University of California, Berkeley, and an email address connecting me to this institutional setting. The politics of access here should not be overlooked, although I have admittedly chosen to underplay them in the core chapters of this text both for narrative reasons and in order to foreground the research and perspectives of those I studied over my own experience.

2 Greenpeace began working in mainland China around 2002, and it established offices in Guangzhou and Beijing shortly afterward. At the time of my interview, the Beijing office was the base of operations for Greenpeace China and fell under the umbrella of both Greenpeace East Asia and Greenpeace International.

3 For example, see Bao et al. (2010).

Chapter 3: The Dietary Environment

1 Their "discovery" through the institutionalized sciences is rooted in agricultural animals falling infertile after, for instance, grazing on red clover (Farnsworth et al. 1975).

2 Important exceptions to this interpretation of S-EDCs include Ann Pollock's "Queering Endocrine Disruption" as well as Nick Shapiro and Eben Kirksey's "Chemo-Ethnography" (Pollock 2016; Shapiro and Kirksey 2017).

3 Usually translated as the cultivation or nurturance of life or health, *yangsheng* "encompasses just about everything one can do to improve one's health, including what tonic to take, what to eat and drink, how to care of one's body, how to relate to time and space and how to relate to other people and the environment" (Sun 2015, 286). Contemporary *yangsheng* practices blend long-standing traditions of caring for the body through food and herbs with more recently developed self-cultivation practices, often accomplished through consumption and promoted through government campaigns (Farquhar and Zhang 2012).

4 Most SNPs are thought to have no impact on health; simply having polymorphisms is not indicative of any specific effect. However, SNPs are also used to associate a likelihood of different traits with individuals and groups.

Chapter 4: The Maternal Environment

An earlier version of this chapter was published in 2016 as "What if the Environment Is a Person? Lineages of Epigenetic Science in a Toxic China" in *Cultural Anthropology* 31 (2).

Chapter 5: The Laboratory Environment

1 This gender imbalance was explained through two premises. First, I was told by PhD candidates that "many graduates of the toxicology program cherished the stable and well-paid appointments in regional CDCs—especially women." Relatedly, I was told a joke, which I had now been told many times during my stays in China, each time by a man and more than once after being asked if I was married: "In China there are three sexes: males, females, and female PhD." (*Shijie shang zhi you san zhong ren, nanren, nuren, nuboshi*). To pursue a PhD in China was to leave one's female gender behind and take on a new identity as an undesirable, unmarriable third sex, according to the connotations of the joke. Indeed, I had seen such dynamics play out in the DeTox Lab. Only one of the women working in there at the time of my research was pursuing her PhD. The remainder of the PhD candidates, postdocs, and junior professors were men.

2 As it is understood and described in scientific texts today, RNA comes in various forms, classified by the taxonomy of superfamily and family. Coding and noncoding superfamilies are defined by their intermediary relationship to DNA. All RNA is transcribed by DNA, but only certain coding RNA also translates DNA into proteins (thus aiding in the transmission of the DNA code). Noncoding RNA (ncRNA) does not translate into proteins, but it is not without purpose, as was once assumed. The ncRNA superfamily has been deemed particularly important and has become increasingly researched. It is now understood that ncRNA regulates DNA expression at transcriptional and post-transcriptional levels. In the words of anthropologist Jörg Niewöhner, "Recent research is beginning to focus on RNA biology and on trans-effects, that is, effects that are mediated via extra-nuclear routes and agents. While not new in principle, this adds another layer of complexity to the field and it is entirely unclear, whether the puzzle of trans-generational stability will be solved at the level of (epigenetic) mechanism any time soon" (2015, 223). With a name that carries both the history of interpreting genes as codes and the move beyond this paradigm toward more-complex theories of post-genomic processes, ncRNA is now considered a part of the crucial mechanics behind genetic expression, with various families of ncRNA involved in these mechanisms, including messenger RNA (mRNA), short and long noncoding RNA (sncRNA and lncRNA), and micro RNA (miRNA).

Epilogue

1 Even today as I write this, while my now young child sleeps in the other room, multiple headlines have found their way into my morning news feed: "The Harmful Chemical Lurking in Your Children's Toys," published in the *New York Times* on November 23, 2020, followed by a suggested reading from August of the same year: "This Chemical Can Impair Fertility, but It's Hard to Avoid."

REFERENCES

Adams, Vincanne, Kathleen Erwin, and Phuoc V. Le. 2010. "Governing through Blood: Biology, Donation, and Exchange in Urban China." In *Asian Biotech: Ethics and Communities of Fate*, edited by Aihwa Ong and Nancy N. Chen, 167–89. Durham, NC: Duke University Press.

Adeoya-Osiguwa, S. A., S. Markoulaki, V. Pocock, S. R. Milligan, and L. R. Fraser. 2003. "17β-Estradiol and Environmental Estrogens Significantly Affect Mammalian Sperm Function." *Human Reproduction* 18, no. 1: 100–107.

Agard-Jones, Vanessa. 2016. *Cultures of Energy Podcast*. Episode 35, September 29. https://cenhs.libsyn.com/ep-35-vanessa-agard-jones.

Ah-King, Malin, and Eva Hayward. 2013. "Toxic Sexes: Perverting Pollution and Queering Hormone Disruption." *O-Zone: A Journal of Object-Oriented Studies*, no. 1: 1–14.

Almeling, Rene. 2011. *Sex Cells: The Medical Market for Eggs and Sperm*. Berkeley: University of California Press.

Almeling, Rene. 2020. *GUYnecology: The Missing Science of Men's Reproductive Health*. Berkeley: University of California Press.

Almeling, Rene, and Miranda R. Waggoner. 2013. "More and Less Than Equal: How Men Factor in the Reproductive Equation." *Gender & Society* 27, no. 6: 821–42.

Anagnost, Ann. 2004. "The Corporeal Politics of Quality (Suzhi)." *Public Culture* 16, no. 2: 189–208.

Atanassova, N., C. McKinnell, K. J. Turner, M. Walker, J. S. Fisher, M. Morley, M. R. Millar, N. P. Groome, and R. M. Sharpe. 2000. "Comparative Effects of Neonatal Exposure of Male Rats to Potent and Weak (Environmental) Estrogens on

Spermatogenesis at Puberty and the Relationship to Adult Testis Size and Fertility: Evidence for Stimulatory Effects of Low Estrogen Levels." *Endocrinology* 141, no. 10: 3898–3907.

Balayannis, Angeliki, and Emma Garnett. 2020. "Chemical Kinship: Interdisciplinary Experiments with Pollution." *Catalyst: Feminism, Theory, Technoscience* 6, no. 1: 1–10.

Bao Chang. 2010. "Infant Formula Scare Spares Dairy Industry." *China Daily*, August 10. http://www.chinadaily.com.cn/business/2010-08/10/content_11129399.htm.

Bao, Jia, Wei Liu, Li Liu, Yihe Jin, Xiaorong Ran, and Zhixu Zhang. 2010. "Perfluorinated Compounds in Urban River Sediments from Guangzhou and Shanghai of China." *Chemosphere* 80, no. 2: 123–30.

Barnes, Liberty Walther. 2014. *Conceiving Masculinity: Male Infertility, Medicine, and Identity*. Philadelphia: Temple University Press.

Becker, Gay. 2000. *The Elusive Embryo: How Women and Men Approach New Reproductive Technologies*. Berkeley: University of California Press.

Benjamin, Ruha. 2018. "Black Afterlives Matter: Cultivating Kinfulness as Reproductive Justice." In *Making Kin Not Population: Reconceiving Generations*, edited by Adele Clarke and Donna J. Haraway, 41–66. Chicago: Prickly Paradigm.

Bernhardt, Emily S., Emma J. Rosi, and Mark O. Gessner. 2017. "Synthetic Chemicals as Agents of Global Change." *Frontiers in Ecology and the Environment* 15, no. 2: 84–90.

Bharadwaj, Aditya. 2013. "Subaltern Biology? Local Biologies, Indian Odysseys, and the Pursuit of Human Embryonic Stem Cell Therapies." *Medical Anthropology* 32, no. 4: 359–73.

Bi, Xiao Zhe. 2010. "*Mei Ge Ren Dou Keneng Chengwei 'ChangJiang Du Yu*" [Every Person Could Become "Poison Yangtze River Fish"]. *Shenyang Ribao*, August 31.

Bohme, Susanna Rankin. 2015. *Toxic Injustice: A Transnational History of Exposure and Struggle*. Berkeley: University of California Press.

Boudia, Soraya, and Nathalie Jas, eds. 2014. *Powerless Science? Science and Politics in a Toxic World*. New York: Berghahn.

Brace, C. Loring. 2005. *"Race" Is a Four-Letter Word: The Genesis of the Concept*. New York: Oxford University Press.

Briggs, Laura. 2018. *How All Politics Became Reproductive Politics: From Welfare Reform to Foreclosure to Trump*. Berkeley: University of California Press.

Carey, Nessa. 2012. *The Epigenetics Revolution: How Modern Biology Is Rewriting Our Understanding of Genetics, Disease, and Inheritance*. New York: Columbia University Press.

Cariati, Federica, Nadja D'Uonno, Francesca Borrillo, Stefania Iervolino, Giacomo Galdiero, and Rossella Tomaiuolo. 2019. "Bisphenol A: An Emerging Threat to Male Fertility." *Reproductive Biology and Endocrinology* 17 (January): 1–8.

Carlsen, Elisabeth, Aleksander Giwercman, Niels Keiding, and Niels E. Skakkebæk. 1992. "Evidence for Decreasing Quality of Semen during Past 50 Years." *British Medical Journal (Clinical Research Ed.)* 305, no. 6854: 609–13.

Carsten, Janet. 2016. *Janet Carsten on the Kinship of Anthropology*. *Social Science Bites*, January. www.socialsciencespace.com/2016/01/janet-carsten-on-the-kinship-of-anthropology.

Casper, Monica J., ed. 2003. *Synthetic Planet: Chemical Politics and the Hazards of Modern Life*. New York: Routledge.

Chang, Chia-ju. 2019. "Environing at the Margins: Huanjing as a Critical Practice." In *Chinese Environmental Humanities: Practices of Environing at the Margins*, edited by Chia-ju Chang, 1–32. Cham: Palgrave Macmillan.

Chao, Deng. 2013. "Newest Pollution Concern: 'Ugly' Sperm." *Wall Street Journal*, November 7. https://blogs.wsj.com/chinarealtime/2013/11/07/chinas-newest-pollution-concern-ugly-sperm.

Chavarro, Jorge E., Thomas L. Toth, Sonita M. Sadio, and Russ Hauser. 2008. "Soy Food and Isoflavone Intake in Relation to Semen Quality Parameters among Men from an Infertility Clinic." *Human Reproduction* 23, no. 11: 2584–90.

Chen, Haidan. 2013. "Governing International Biobank Collaboration: A Case Study of China Kadoorie Biobank." *Science, Technology and Society* 18, no. 3: 321–38.

Chen, Liyu. 2013. "*Shi Nianlai 2/3 Juan Jing Zhe Jingzi Huoli Bu Dabiao*" [Over the Last 10 Years 2/3 of Sperm Donors' Sperm Vitality Does Not Reach Standards]. *Shanghai Morning Post*, November 6. data.jfdaily.com/a/7178924.htm.

Chen, Mel. 2012. *Animacies: Biopolitics, Racial Mattering, and Queer Affect.* Durham, NC: Duke University Press.

Chen, Nancy N. 2003. *Breathing Spaces: Qigong, Psychiatry, and Healing in China.* New York: Columbia University Press.

Chen, Nancy N. 2010. "Feeding the Nation: Chinese Biotechnology and Genetically Modified Foods." In *Asian Biotech: Ethics and Communities of Fate*, edited by Aihwa Ong and Nancy N. Chen, 81–92. Durham, NC: Duke University Press.

Chen, Zhu, and Guo-ping Zhao. 2009. "Human Genomics in China—Ten Years Endeavor: From Planning to Implementation." *Science in China Series C: Life Sciences* 52, no. 1: 2–6.

Cheng, Yuanyuan, and C. Paul Nathanail. 2019. "A Study of 'Cancer Villages' in Jiangsu Province of China." *Environmental Science and Pollution Research International* 26, no. 2: 1932–46. https://doi.org/10.1007/s11356-018-3758-4.

Chin, Tai-Wai, and Chih-Yang Chiu. 2008. "Analysis of the RET Gene in Subjects with Sporadic Hirschsprung's Disease." *Journal of the Chinese Medical Association* 71, no. 8: 406–10.

"China Investigates Claims Tainted Milk Powder Made Infant Girls 'Grow Breasts.'" 2010. *Guardian*, August 10. www.theguardian.com/world/2010/aug/10/china-milk-powder-girls-breasts.

Choy, Timothy K. 2011. *Ecologies of Comparison: An Ethnography of Endangerment in Hong Kong.* Durham, NC: Duke University Press.

Clarke, Adele E., Janet K. Shim, Laura Mamo, Jennifer Ruth Fosket, and Jennifer R. Fishman. 2003. "Biomedicalization: Technoscientific Transformations of Health, Illness, and U.S. Biomedicine." *American Sociological Review* 68, no. 2: 161–94. https://doi.org/10.2307/1519765.

Clarke, Adele, and Donna J. Haraway, eds. 2018. *Making Kin Not Population: Reconceiving Generations.* Chicago: Prickly Paradigm.

Cohen, Lawrence. 2020. "The Culling: Pandemic, Gerocide, Generational Affect." *Medical Anthropology Quarterly* 34, no. 4: 542–60. https://doi.org/10.1111/maq.12627.

Colborn, Theo, John Peterson Myers, and Dianne Dumanoski. 1996. *Our Stolen Future: How We Are Threatening Our Fertility, Intelligence, and Survival: A Scientific Detective Story*. New York: Dutton.

Collier, Jane F., and Sylvia J. Yanagisako. 1989. "Theory in Anthropology since Feminist Practice." *Critique of Anthropology* 9, no. 2: 27–37.

Daniels, Cynthia R. 2006. *Exposing Men: The Science and Politics of Male Reproduction*. Oxford: Oxford University Press.

Darling, Katherine Weatherford, Sara L. Ackerman, Robert H. Hiatt, Sandra Soo-Jin Lee, and Janet K. Shim. 2016. "Enacting the Molecular Imperative: How Gene-Environment Interaction Research Links Bodies and Environments in the Post-Genomic Age." *Social Science and Medicine* 155 (April): 51–60. https://doi.org/10.1016/j.socscimed.2016.03.007.

Davis, Heather. 2015. "Toxic Progeny: The Plastisphere and Other Queer Futures." *PhiloSOPHIA* 5, no. 2: 231–50.

Deans, Carrie, and Keith A. Maggert. 2015. "What Do You Mean, 'Epigenetic?'" *Genetics* 199, no. 4: 887–96.

Di Chiro, Giovanna. 2010. "Polluted Politics? Confronting Toxic Discourse, Sex Panic, and Eco-Normativity." In *Queer Ecologies: Sex, Nature, Politics, Desire*, edited by Catriona Mortimer-Sandilands and Bruce Erickson, 199–230. Bloomington: Indiana University Press.

Dikötter, Frank. 1998. *Imperfect Conceptions: Medical Knowledge, Birth Defects, and Eugenics in China*. New York: Columbia University Press.

Dow, Katharine. 2016. *Making a Good Life: An Ethnography of Nature, Ethics, and Reproduction*. Princeton, NJ: Princeton University Press.

DuPuis, E. Melanie. 2007. "Angels and Vegetables: A Brief History of Food Advice in America." *Gastronomica* 7, no. 3: 34–44.

Duster, Troy. 2005. "Race and Reification in Science." *Science* 307, no. 5712: 1050–51.

Edelman, Lee. 2004. *No Future: Queer Theory and the Death Drive*. Durham, NC: Duke University Press.

Emison, Eileen Sproat, Andrew S. McCallion, Carl S. Kashuk, Richard T. Bush, Elizabeth Grice, Shin Lin, Matthew E. Portnoy, David J. Cutler, Eric D. Green, and Aravinda Chakravarti. 2005. "A Common Sex-Dependent Mutation in a RET Enhancer Underlies Hirschsprung Disease Risk." *Nature* 434, no. 7035: 857–63.

Ericson, Anders, and Bengt Källén. 2001. "Congenital Malformations in Infants Born after IVF: A Population-Based Study." *Human Reproduction* 16, no. 3: 504–9.

Fabian, Ann. 2010. *The Skull Collectors: Race, Science, and America's Unburied Dead*. Chicago: University of Chicago Press.

Faircloth, Charlotte. 2014. "Intensive Parenting and the Expansion of Parenting." In *Parenting Culture Studies*, 25–50. New York: Palgrave Macmillan.

Farnsworth, Norman R., Audrey S. Bingel, Geoffrey A. Cordell, Frank A. Crane, and Harry H. S. Fong. 1975. "Potential Value of Plants as Sources of New Antifertility Agents II." *Journal of Pharmaceutical Sciences* 64, no. 5: 717–54.

Farquhar, Judith. 1991. "Objects, Processes, and Female Infertility in Chinese Medicine." *Medical Anthropology Quarterly* 5, no. 4: 370–99.

Farquhar, Judith. 2002. *Appetites: Food and Sex in Postsocialist China*. Durham, NC: Duke University Press.

Farquhar, Judith, and Qicheng Zhang. 2012. *Ten Thousand Things: Nurturing Life in Contemporary Beijing*. New York: Zone.

Fausto-Sterling, Anne. 2000. *Sexing the Body: Gender Politics and the Construction of Sexuality*. New York: Basic.

Fearnley, Lyle. 2013. "Limn: The Birds of Poyang Lake: Sentinels at the Interface of Wild and Domestic." In "Sentinel Devices," edited by Frédéric Keck and Andrew Lakoff, special issue, *Limn* 3 (June). https://limn.it/articles/the-birds-of-poyang-lake-sentinels-at-the-interface-of-wild-and-domestic.

Fei, Xiaotong. 1992. *From the Soil: The Foundations of Chinese Society*. Translated by Gary G. Hamilton and Wang Zheng. Berkeley: University of California Press.

Floehr, Tilman, Hongxia Xiao, Björn Scholz-Starke, Lingling Wu, Junli Hou, Daqiang Yin, Xiaowei Zhang, et al. 2013. "Solution by Dilution?—A Review on the Pollution Status of the Yangtze River." *Environmental Science and Pollution Research* 20, no. 10: 6934–71.

Fortun, Kim. 2011. "Toxics Trouble: Feminism and the Subversion of Science." In *Körper Raum Transformation*, edited by Elvira Scheich and Karen Wagels, 234–54. Münster: Westfälisches Dampfboot.

Fortun, Kim. 2012. "Ethnography in Late Industrialism." *Cultural Anthropology* 27, no. 3: 446–64.

Fortun, Kim, and Mike Fortun. 2005. "Scientific Imaginaries and Ethical Plateaus in Contemporary U.S. Toxicology." *American Anthropologist* 107, no. 1: 43–54.

Franklin, Sarah. 1988. "Life Story: The Gene as Fetish Object on TV." *Science as Culture* 1, no. 3: 92–100.

Franklin, Sarah. 1997. *Embodied Progress: A Cultural Account of Assisted Conception*. London: Routledge.

Franklin, Sarah. 2000. "Life Itself: Global Nature and the Genetic Imaginary." In *Global Nature, Global Culture*, edited by Sarah Franklin, Celia Lury, and Jackie Stacey, 188–227. London: Sage.

Franklin, Sarah. 2005. "Stem Cells R Us: Emergent Life Forms and the Global Biological." In *Global Assemblages: Technology, Politics, and Ethics as Anthropological Problems*, edited by Aihwa Ong and Stephen J. Collier, 59–78. Malden, MA: Blackwell.

Franklin, Sarah. 2007. "Obituary: Dame Dr Anne McLaren." *Regenerative Medicine* 2, no. 5: 853–59.

Franklin, Sarah. 2011. "Transbiology: A Feminist Cultural Account of Being after IVF." *The Scholar & Feminist Online* 9, no. 1. http://sfonline.barnard.edu/reprotech/franklin_01.htm.

Franklin, Sarah. 2013a. *Biological Relatives: IVF, Stem Cells, and the Future of Kinship*. Durham, NC: Duke University Press.

Franklin, Sarah. 2013b. "From Blood to Genes? Rethinking Consanguinity in the Context of Geneticization." In *Blood and Kinship: Matter for Metaphor from Ancient Rome to the Present*, edited by Christopher H. Johnson, Bernhard Jussen, David Warren Sabean, and Simon Teuscher, 285–306. New York: Berghahn.

Fraser, Lynn R., Ergin Beyret, Stuart R. Milligan, and Susan A. Adeoya-Osiguwa. 2006. "Effects of Estrogenic Xenobiotics on Human and Mouse Spermatozoa." *Human Reproduction* 21, no. 5: 1184–93.

Frickel, Scott. 2004. *Chemical Consequences: Environmental Mutagens, Scientist Activism, and the Rise of Genetic Toxicology*. New Brunswick, NJ: Rutgers University Press.

Friese, Carrie, Gay Becker, and Robert D. Nachtigall. 2006. "Rethinking the Biological Clock: Eleventh-Hour Moms, Miracle Moms and Meanings of Age-Related Infertility." *Social Science & Medicine* 63, no. 6: 1550–60.

Friese, Carrie, and Joanna Latimer. 2019. "Entanglements in Health and Well-Being: Working with Model Organisms in Biomedicine and Bioscience." *Medical Anthropology Quarterly* 33, no. 1: 120–37.

Fu, Jia-Chen. 2017. "Artemisinin and Chinese Medicine as Tu Science." *Endeavour* 41, no. 3: 127–35.

Fu, Jia-Chen. 2018. *The Other Milk: Reinventing Soy in Republican China*. Seattle: University of Washington Press.

Fuentes, Agustín. 2015. *Race, Monogamy, and Other Lies They Told You: Busting Myths about Human Nature*. Berkeley: University of California Press.

Fullwiley, Duana. 2007. "The Molecularization of Race: Institutionalizing Human Difference in Pharmacogenetics Practice." *Science as Culture* 16, no. 1: 1–30.

Fullwiley, Duana. 2012. *The Enculturated Gene: Sickle Cell Health Politics and Biological Difference in West Africa*. Princeton, NJ: Princeton University Press.

Gambert, Iselin, and Tobias Linné. 2018. "From Rice Eaters to Soy Boys: Race, Gender and Tropes of 'Plant Food Masculinity.'" December 9. https://ssrn.com/abstract=3298467.

Ghosh, Pallab. 2017. "Sperm Count Drop 'Could Make Humans Extinct.'" BBC News, July 25. www.bbc.com/news/health-40719743.

Gibbon, Sahra, and Michelle Pentecost. 2019. "Introduction: Excavating and (Re) Creating the Biosocial: Birth Cohorts as Ethnographic Object of Inquiry and Site of Intervention." *Somatosphere*, November 15. http://somatosphere.net/2019/introduction-excavating-and-recreating-the-biosocial-birth-cohorts-as-ethnographic-object-of-inquiry-and-site-of-intervention.html.

Gibbon, Sahra, Barbara Prainsack, Stephen Hilgartner, and Janelle Lamoreaux, eds. 2018. *Routledge Handbook of Genomics, Health and Society*. 2nd ed. Abingdon: Routledge.

Gibbon, Sahra, Barbara Prainsack, Stephen Hilgartner, and Janelle Lamoreaux. 2018. "Introduction to Handbook of Genomics, Health and Society." In *Routledge Handbook of Genomics, Health and Society*, 2nd ed., edited by Sahra Gibbon, Barbara Prainsack, Stephen Hilgartner, and Janelle Lamoreaux, 1–8. Abingdon: Routledge.

Ginsburg, Faye, and Rayna Rapp. 1991. "The Politics of Reproduction." *Annual Review of Anthropology* 20: 311–43.

Ginsburg, Faye, and Rayna Rapp. 2019. "Disability/Anthropology: Rethinking the Parameters of the Human." *Current Anthropology* 61 (Supplement 21): S4–S15. https://doi.org/10.1086/705503.

Giudice, Linda C. 2016. "Environmental Toxicants: Hidden Players on the Reproductive Stage." *Fertility and Sterility* 106, no. 4: 791–94. https://doi.org/10.1016/j.fertnstert.2016.08.019.

Gordon, Deborah. 1988. "Tenacious Assumptions of Western Medicine." In *Biomedicine Examined*, edited by Margaret Lock and Deborah Gordon, 19–56. Springer Nature.

Goron, Coraline. 2018. "Ecological Civilisation and the Political Limits of a Chinese Concept of Sustainability." *China Perspectives* 4: 39–52.

Graeber, David. 2009. *Direct Action: An Ethnography*. Edinburgh: AK Press.

Greenhalgh, Susan. 1994. "Controlling Births and Bodies in Village China." *American Ethnologist* 21, no. 1: 3–30.

Greenhalgh, Susan. 2009. "The Chinese Biopolitical: Facing the Twenty-First Century." *New Genetics and Society* 28, no. 3: 205–22.

Greenhalgh, Susan. 2010. *Cultivating Global Citizens: Population in the Rise of China*. Cambridge, MA: Harvard University Press.

Greenhalgh, Susan. 2013. "Patriarchal Demographics? China's Sex Ratio Reconsidered." *Population and Development Review* 38, no. s1: 130–49.

Greenhalgh, Susan, and Edwin A. Winckler. 2005. *Governing China's Population: From Leninist to Neoliberal Biopolitics*. Stanford, CA: Stanford University Press.

Greenhalgh, Susan, and Li Zhang, eds. 2020. *Can Science and Technology Save China?* Ithaca, NY: Cornell University Press.

Greenpeace. 2010. *Swimming in Poison: An Analysis of Hazardous Chemicals in Yangtze River Fish*. Beijing: Greenpeace.

Guthman, Julie, and Becky Mansfield. 2013. "The Implications of Environmental Epigenetics: A New Direction for Geographic Inquiry on Health, Space, and Nature-Society Relations." *Progress in Human Geography* 37, no. 4: 486–504.

Hamburger, Jessica. 2002. "Pesticides in China: A Growing Threat to Food Safety, Public Health, and the Environment." In *China Environment Series* 5, 29–44. Washington, DC: Wilson Center. www.wilsoncenter.org/publication/china-environment-series-5-2002.

Han, Zhe. 2010. "Who Will Give 'Hormone-Ridden' Wild Yangtze River Fish a Voice?" [*Shei Gei "Beijisu" de Yesheng Changjiang Yu Yige Shuofa?*]. *Beijing Business News (Beijing Shangbao)*, September 2.

Handwerker, Lisa. 2002. "The Politics of Making Modern Babies in China." In *Infertility around the Globe: New Thinking on Childlessness, Gender, and Reproductive Technologies*, edited by Marcia C. Inhorn and Frank Van Balen, 298–315. Berkeley: University of California Press.

Haraway, Donna J. 1988. "Situated Knowledges: The Science Question in Feminism and the Privilege of Partial Perspective." *Feminist Studies* 14, no. 3: 575–99. https://doi.org/10.2307/3178066.

Haraway, Donna J. 1991. *Simians, Cyborgs, and Women: The Reinvention of Nature*. New York: Routledge.

Haraway, Donna J. 1997a. *Modest_Witness@Second_Millennium. FemaleMan©_Meets_OncoMouse™: Feminism and Technoscience*. New York: Routledge.

Haraway, Donna J. 1997b. "The Virtual Speculum in the New World Order." *Feminist Review* 55, 1, 22–72.

Haraway, Donna J. 2008. *When Species Meet*. Minneapolis: University of Minnesota Press.

Hayden, Cori. 1995. "Gender, Genetics, and Generation: Reformulating Biology in Lesbian Kinship." *Cultural Anthropology* 10, no. 1: 41–63.

Hayden, Cori. 2012. "Rethinking Reductionism, or, the Transformative Work of Making the Same." *Anthropological Forum* 22, no. 3: 271–83.

Hedlund, Maria. 2011. "Epigenetic Responsibility." *Medicine Studies* 3, no. 3: 171–83.

Henderson, Rogene F., William E. Bechtold, James A. Bond, and James D. Sun. 1989. "The Use of Biological Markers in Toxicology." *Critical Reviews in Toxicology* 20, no. 2: 65–82.

Hoffman, Lisa. 2006. "Governing through Environment (Huanjing): Sustainability, Value, and City Building in Dalian, China." Paper presented at the 3rd Annual China Planning Network Conference, Beijing, China, June 14. http://web.mit.edu/dusp/chinaplanning/paper/Hoffman%20paper.pdf.

Holdaway, Jennifer. 2013. "Environment and Health Research in China: The State of the Field." *China Quarterly* 214: 255–82.

Hollert, H. 2013. "Processes and Environmental Quality in the Yangtze River System." *Environmental Science and Pollution Research* 20, no. 10: 6904–6. http://dx.doi.org.ezproxy3.library.arizona.edu/10.1007/s11356-013-1943-z.

Hong, Xin Liang. 2010. "Yangtze River Fish Poisoned, Administrators Shouldn't Wait" [*Changjiang Yu Youdu, Zhili Buneng Zai Deng*]. *Qianjiang Evening News (Qianjang Wanbao)*, August 31.

Hoover, Elizabeth. 2017. *The River Is in Us: Fighting Toxics in a Mohawk Community.* Minneapolis: University of Minnesota Press.

Horton, Sarah, and Judith C Barker. 2010. "Stigmatized Biologies: Examining the Cumulative Effects of Oral Health Disparities for Mexican American Farmworker Children." *Medical Anthropology Quarterly* 24, no. 2: 199–219.

Hsu, Elisabeth. 2009. "Diverse Biologies and Experiential Continuities: Did the Ancient Chinese Know That Qinghao Had Anti-malarial Properties?" *Canadian Bulletin of Medical History* 26, no. 1: 203–14.

Huang, Hsing-Tsung. 2008. "Early Uses of Soybean in Chinese History." In *The World of Soy*, edited by Christine M. Du Bois, Chee-Beng Tan, and Sydney W. Mintz, 45–55. Urbana: University of Illinois Press.

Hubbert, Jennifer. 2015. "'We're Not THAT Kind of Developing Country': Environmental Awareness in Contemporary China." In *Sustainability in the Global City: Myth and Practice*, edited by Cindy Isenhour, Gary McDonogh, and Melissa Checker, 29–53. New York: Cambridge University Press.

Inhorn, M. C., D. Birenbaum-Carmeli, J. Birger, L. M. Westphal, J. Doyle, N. Gleicher, D. Meirow, et al. 2018. "Elective Egg Freezing and Its Underlying Sociodemography: A Binational Analysis with Global Implications." *Reproductive Biology and Endocrinology*: 16, no. 70: 1–11.

"International Greenpeace Organization Publishes 'Poison' Hidden in River." 2010. *Beijing Shangbao (Beijing Business News)*, August 30.

Jablonka, Eva, and Marion J. Lamb. 2006. *Evolution in Four Dimensions: Genetic, Epigenetic, Behavioral, and Symbolic Variation in the History of Life*. Cambridge, MA: MIT Press.

Jiang, Lijing. 2015. "IVF the Chinese Way: Zhang Lizhu and Post-Mao Human In Vitro Fertilization Research." *East Asian Science, Technology and Society* 9, no. 1: 23–45.

Jiang, Lijing. 2017. "Crafting Socialist Embryology: Dialectics, Aquaculture and the Diverging Discipline in Maoist China, 1950–1965." *History and Philosophy of the Life Sciences* 40, no. 1: 3.

Jing, Jun. 2000. "Introduction." In *Feeding China's Little Emperors: Food, Children, and Social Change,* edited by Jun Jing, 1–26. Stanford, CA: Stanford University Press.

Jinhuozaifendou. 2013. Reply to "China's Newest Pollution Concern: 'Ugly' Sperm" [*Zhongguoren Dui Wuran de Zuixin Danyou: "Nankan" de jingzi*]. *Wall Street Journal Chinese (Huaerjie Ribao Zhongwenwang),* November 8. www.cn.wsj.com/gb/20131108/ren100634asp.

Johnson, Larry, Jeffrey J. Barnard, Lori Rodriguez, Elizabeth C. Smith, Ronald S. Swerdloff, X. H. Wang, and Christina Wang. 1998. "Ethnic Differences in Testicular Structure and Spermatogenic Potential May Predispose Testes of Asian Men to a Heightened Sensitivity to Steroidal Contraceptives." *Journal of Andrology* 19, no. 3: 348–57.

Kafer, Alison. 2019. "Crip Kin, Manifesting." *Catalyst: Feminism, Theory, Technoscience* 5, no. 1: 1–37.

Keck, Frédéric, and Andrew Lakoff. 2013. "Preface: Sentinel Devices." *Limn* (May). http://limn.it/preface-sentinel-devices-2.

Keller, Evelyn Fox. 1984. *A Feeling for the Organism: The Life and Work of Barbara McClintock.* New York: Henry Holt.

Keller, Evelyn Fox. 2002. *The Century of the Gene.* Cambridge, MA: Harvard University Press.

Kier, Bailey. 2010. "Interdependent Ecological Transsex: Notes on Re/Production, 'Transgender' Fish, and the Management of Populations, Species, and Resources." *Women & Performance: A Journal of Feminist Theory* 20, no. 3: 299–319. https://doi.org/10.1080/0740770X.2010.529254.

Kimura, Aya Hirata. 2016. *Radiation Brain Moms and Citizen Scientists: The Gender Politics of Food Contamination after Fukushima.* Durham, NC: Duke University Press.

kingarthurzj_9006. 2013. Reply to "China's Newest Pollution Concern: 'Ugly' Sperm." [*Zhongguoren Dui Wuran de Zuixin Danyou: "Nankan" de jingzi*]. *Wall Street Journal Chinese (Huaerjie Ribao Zhongwenwang),* November 8. www.cn.wsj.com/gb/20131108/ren100634asp.

Kleinman, Arthur. 1981. *Patients and Healers in the Context of Culture: An Exploration of the Borderland between Anthropology, Medicine, and Psychiatry.* Berkeley: University of California Press.

Knorr-Cetina, Karin. 1999. *Epistemic Cultures: How the Sciences Make Knowledge.* Cambridge, MA: Harvard University Press.

Kohrman, Matthew. 2005. *Bodies of Difference: Experiences of Disability and Institutional Advocacy in the Making of Modern China.* Berkeley: University of California Press.

Kohrman, Matthew. 2020. "Unmasking a Gendered Materialism: Air Filtration, Cigarettes and Domestic Discord in Urban China." In *Can Science and Technology Save China?,* edited by Susan Greenhalgh and Li Zhang, 184–212. Ithaca, NY: Cornell University Press.

Krimsky, Sheldon. 2000. *Hormonal Chaos: The Scientific and Social Origins of the Environmental Endocrine Hypothesis*. Baltimore, MD: Johns Hopkins University Press.

Kubsad, Deepika, Eric E. Nilsson, Stephanie E. King, Ingrid Sadler-Riggleman, Daniel Beck, and Michael K. Skinner. 2019. "Assessment of Glyphosate Induced Epigenetic Transgenerational Inheritance of Pathologies and Sperm Epimutations: Generational Toxicology." *Scientific Reports* 9, no. 1: 6372.

Kuzawa, Christopher W., and Elizabeth Sweet. 2009. "Epigenetics and the Embodiment of Race: Developmental Origins of US Racial Disparities in Cardiovascular Health." *American Journal of Human Biology* 21, no. 1: 2–15.

Lam, Wai-man. 2014. "Nongovernmental International Human Rights Organizations: The Case of Hong Kong." *PS: Political Science and Politics* 47, no. 3: 642–53.

Lamoreaux, Janelle, and Ayo Wahlberg. 2022. "Sperm." In *An Anthropogenic Table of Elements*, edited by Courtney Addison, Timothy Neale, and Thao Phan, 158–67. Toronto, Canada: University of Toronto Press.

Landecker, Hannah. 2011. "Food as Exposure: Nutritional Epigenetics and the New Metabolism." *BioSocieties* 6, no. 2: 167–94.

Landecker, Hannah. 2013. "Postindustrial Metabolism: Fat Knowledge." *Public Culture* 25, no. 3 (71): 495–522.

Landecker, Hannah. 2014. "Pregnancy: Study the Mother's DNA as Well." *Nature* 513, no. 7517: 172–72.

Landecker, Hannah. 2016. "The Social as Signal in the Body of Chromatin." *Sociological Review* 64, no. 1_suppl: 79–99.

Landecker, Hannah, and Aaron Panofsky. 2013. "From Social Structure to Gene Regulation, and Back: A Critical Introduction to Environmental Epigenetics for Sociology." *Annual Review of Sociology* 39, no. 1: 333–57.

Langston, Nancy. 2011. *Toxic Bodies: Hormone Disruptors and the Legacy of DES*. New Haven, CT: Yale University Press.

Lappé, Martine. 2016. "The Maternal Body as Environment in Autism Science." *Social Studies of Science* 46, no. 5: 675–700.

Lappé, Martine. 2018. "The Paradox of Care in Behavioral Epigenetics: Constructing Early-Life Adversity in the Lab." *BioSocieties* 13, no. 4: 698–714.

Latour, Bruno. 1987. *Science in Action: How to Follow Scientists and Engineers through Society*. Cambridge, MA: Harvard University Press.

Latour, Bruno, and Steve Woolgar. 1986. *Laboratory Life*. Princeton, NJ: Princeton University Press.

Law, John, and Wen-yuan Lin. 2017. "Provincializing STS: Postcoloniality, Symmetry, and Method." *East Asian Science, Technology and Society* 11, no. 2: 211–27.

Li, Jialin. 2021. "Cloaking the Pregnancy: Scientific Uncertainty and Gendered Burden among Middle-Class Mothers in Urban China." *Science, Technology, & Human Values* 46, no. 1: 3–28.

Li, Peishan. 1988. "Genetics in China: The Qingdao Symposium of 1956." *Isis* 79, no. 2: 227–36.

Litzinger, Ralph, and Fan Yang. 2019. "Ecomedia Events in China: From Yellow Eco-peril to Media Materialism." In *Chinese Environmental Humanities: Practices of Environing at the Margins*, edited by Chia-ju Chang, 209–35. Cham: Palgrave Macmillan.

Liu, Xin. 2012. *The Mirage of China*. New York: Berghahn.

Liu, Zhangjing, Sudan Zhang, and Ying Chen. 2010. "Wild Fish in Yangtze River 'Poisoned' Implicating Fish in Xiang River and Dongting Lake" [*Changjiang Yesheng Yu "Han Du" Lianlei Xiangjiang Dongting Yu*]. *Sanxiang City News (Sanxiangdushibao)*, September 13.

Lock, Margaret. 1993. *Encounters with Aging Mythologies of Menopause in Japan and North America*. Berkeley: University of California Press.

Lock, Margaret, and Judith Farquhar. 2007. *Beyond the Body Proper: Reading the Anthropology of Material Life*. Durham, NC: Duke University Press.

Lock, Margaret M., and Gisli Palsson. 2016. *Can Science Resolve the Nature/Nurture Debate?* Malden, MA: Polity.

Lora-Wainwright, Anna. 2017. *Resigned Activism: Living with Pollution in Rural China*. Cambridge, MA: MIT Press.

Luciano, Dana, and Mel Y. Chen. 2015. "Introduction: Has the Queer Ever Been Human?" *GLQ: A Journal of Lesbian and Gay Studies* 21, no. 2: iv–207.

Luo, Hui-Yuan. 1988. "Medical Genetics in China." *Journal of Medical Genetics* 25, no. 4: 253–57.

Lysenko, T. D. [1951] 2001. *Heredity and Its Variability*. Honolulu, HI: University Press of the Pacific.

MacKendrick, Norah. 2014. "More Work for Mother: Chemical Body Burdens as a Maternal Responsibility." *Gender & Society* 28, no. 5: 705–28.

MacKendrick, Norah. 2018. *Better Safe Than Sorry: How Consumers Navigate Exposure to Everyday Toxics*. Berkeley: University of California Press.

Makinen, Julie. 2016. "Hundreds of Chinese Kids Fall Ill from Toxic Chemicals at School, State TV Reports." *Los Angeles Times*, April 19. www.latimes.com/world/asia/la-fg-china-school-toxic-20160419-story.html.

Mansfield, Becky. 2012. "Race and the New Epigenetic Biopolitics of Environmental Health." *BioSocieties* 7, no. 4: 352–72.

Mansfield, Becky. 2017. "Folded Futurity: Epigenetic Plasticity, Temporality, and New Thresholds of Fetal Life." *Science as Culture* 26, no. 3: 355–79.

Mansfield, Becky, and Julie Guthman. 2015. "Epigenetic Life: Biological Plasticity, Abnormality, and New Configurations of Race and Reproduction." *Cultural Geographies* 22, no. 1: 3–20.

Martin, Emily. 1987. *The Woman in the Body: A Cultural Analysis of Reproduction*. Boston: Beacon.

Martin, Emily. 1990. "Toward an Anthropology of Immunology: The Body as Nation State." *Medical Anthropology Quarterly* 4, no. 4: 410–26.

Martin, Emily. 1991. "The Egg and the Sperm: How Science Has Constructed a Romance Based on Stereotypical Male-Female Roles." *Signs* 16, no. 3: 485–501.

Martin, Emily. 1994. *Flexible Bodies: Tracking Immunity in American Culture from the Days of Polio to the Age of AIDS*. Boston: Beacon.

Martin, Emily. 2006. "The Pharmaceutical Person." *BioSocieties* 1, no. 3: 273–87.

Masco, Joseph. 2017. "The Crisis in Crisis." *Current Anthropology* 58 (February): S65–S76.

Meloni, Maurizio, and Giuseppe Testa. 2014. "Scrutinizing the Epigenetics Revolution." *Biosocieties* 9, no. 4: 431–56.

Ministry of Health, People's Republic of China. 2011. "Report on Women and Children's Health Development in China." www.gov.cn/gzdt/att/att/site1/20110921/001e3741a4740fe3bdbf02.pdf.

Mol, Annemarie. 2002. *The Body Multiple: Ontology in Medical Practice*. Durham, NC: Duke University Press.

Moore, Lisa Jean. 2007. *Sperm Counts: Overcome by Man's Most Precious Fluid*. New York: New York University Press.

Morgan, Lynn. 2009. *Icons of Life: A Cultural History of Human Embryos*. Berkeley: University of California Press.

Morgan, Lynn M., and Elizabeth F. S. Roberts. 2012. "Reproductive Governance in Latin America." *Anthropology & Medicine* 19, no. 2: 241–54.

Mueggler, Erik. 2001. *The Age of Wild Ghosts: Memory, Violence, and Place in Southwest China*. Berkeley: University of California Press.

Müller, Ruth, Clare Hanson, Mark Hanson, Michael Penkler, Georgia Samaras, Luca Chiapperino, John Dupré, et al. 2017. "The Biosocial Genome?" *EMBO Reports* 18, no. 10: 1677–82.

Murphy, Michelle. 2006. *Sick Building Syndrome and the Problem of Uncertainty: Environmental Politics, Technoscience, and Women Workers*. Durham, NC: Duke University Press.

Murphy, Michelle. 2013. "Distributed Reproduction, Chemical Violence, and Latency." *The Scholar & Feminist Online*. http://sfonline.barnard.edu/life-un-ltd-feminism-bioscience-race/distributed-reproduction-chemical-violence-and-latency.

Nading, Alex M. 2020. "Living in a Toxic World." *Annual Review of Anthropology* 49: 209–24.

Nash, Linda Lorraine. 2006. *Inescapable Ecologies: A History of Environment, Disease, and Knowledge*. Berkeley: University of California Press.

Nelkin, Dorothy, and M. Susan Lindee. 2004. *The DNA Mystique*. Ann Arbor: University of Michigan Press.

Nelson, Nicole C. 2018. *Model Behavior: Animal Experiments, Complexity, and the Genetics of Psychiatric Disorders*. Chicago: University of Chicago Press.

Niewöhner, Jörg. 2015. "Epigenetics: Localizing Biology through Co-laboration." *New Genetics and Society* 34, no. 2: 219–42.

Niewöhner, Jörg, and Margaret Lock. 2018. "Situating Local Biologies: Anthropological Perspectives on Environment/Human Entanglements." *BioSocieties* 13, no. 4: 681–97.

Ong, Aihwa, and Nancy N. Chen, eds. 2010. *Asian Biotech: Ethics and Communities of Fate*. Durham, NC: Duke University Press.

Ong, Aihwa, and Stephen J. Collier, eds. 2004. *Global Assemblages: Technology, Politics, and Ethics as Anthropological Problems*. Malden, MA: Wiley-Blackwell.

Ortner, Sherry B. 1984. "Theory in Anthropology since the Sixties." *Comparative Studies in Society and History* 26, no. 1: 126–66.

Oudshoorn, Nelly. 1994. *Beyond the Natural Body: An Archaeology of Sex Hormones*. New York: Routledge.

Oxfeld, Ellen. 2017. *Bitter and Sweet: Food, Meaning and Modernity in China*. Berkeley: University of California Press.

Pentecost, Michelle, and Maurizio Meloni. 2020. "'It's Never Too Early': Preconception Care and Postgenomic Models of Life." *Frontiers in Sociology* 5: 21.

Pentecost, Michelle, and Fiona Ross. 2019. "The First Thousand Days: Motherhood, Scientific Knowledge, and Local Histories." *Medical Anthropology* 38, no. 8: 747–61.

Perry, Melissa. 2015. "Chemically Induced DNA Damage and Sperm and Oocyte Repair Machinery: The Story Gets More Interesting." *Asian Journal of Andrology* 17, nos. 1–2. www.ajandrology.com/preprintarticle.asp?id=156118.

Petchesky, Rosalind Pollack. 1987. "Fetal Images: The Power of Visual Culture in the Politics of Reproduction." *Feminist Studies* 13, no. 2: 263–92. https://doi.org/10.2307/3177802.

Peterson, Erik L. 2016. *The Life Organic: The Theoretical Biology Club and the Roots of Epigenetics:* Pittsburgh: University of Pittsburgh Press.

Petryna, Adriana. 2002. *Life Exposed: Biological Citizens after Chernobyl.* Princeton, NJ: Princeton University Press.

Phillips, Natalia Ledo Husby, and Tania L. Roth. 2019. "Animal Models and Their Contribution to Our Understanding of the Relationship between Environments, Epigenetic Modifications, and Behavior." *Genes* 10, no. 1: 47.

Pickersgill, Martyn, Jörg Niewöhner, Ruth Müller, Paul Martin, and Sarah Cunningham-Burley. 2013. "Mapping the New Molecular Landscape: Social Dimensions of Epigenetics." *New Genetics and Society* 32, no. 4: 429–47.

Ping, Ping, Wen-Bing Zhu, Xin-Zong Zhang, Yu-Shan Li, Quan-Xian Wang, Xiao-Rong Cao, Yong Liu, Hui-Li Dai, Yi-Ran Huang, and Zheng Li. 2011. "Sperm Donation and Its Application in China: A 7-Year Multicenter Retrospective Study." *Asian Journal of Andrology* 13, no. 4: 644–48.

Pollock, Anne. 2016. "Queering Endocrine Disruption." In *Object-Oriented Feminism*, edited by Katherine Behar, 183–99. Minneapolis: University of Minnesota Press.

Qiao, Jie, and Huai L. Feng. 2014. "Assisted Reproductive Technology in China: Compliance and Non-compliance." *Translational Pediatrics* 3, no. 2: 91–97.

Raffles, Hugh. 2010. "The Quality of Queerness Is Not Strange Enough." In *Insectopedia*, 257–63. New York: Pantheon.

Rapp, Rayna R. 1999. *Testing Women, Testing the Fetus: The Social Impact of Amniocentesis in America.* New York: Routledge.

Rapp, Rayna. 2005. "Comments (on Margaret Lock's Eclipse of the Gene and the Return of Divination)." *Current* 46, no. S5: S64–S65.

Reamon-Buettner, Stella Marie, Vanessa Mutschler, and Juergen Borlak. 2008. "The Next Innovation Cycle in Toxicogenomics: Environmental Epigenetics." *Mutation Research* 659, nos. 1–2: 158–65.

Ren, Hai. 2013. *The Middle Class in Neoliberal China: Governing Risk, Life-Building, and Themed Spaces.* London: Routledge.

Ren, Hai. 2015. *Neoliberalism and Culture in China and Hong Kong: The Countdown of Time.* London: Routledge.

Richardson, Sarah S., Cynthia R. Daniels, Matthew W. Gillman, Janet Golden, Rebecca Kukla, Christopher Kuzawa, and Janet Rich-Edwards. 2014. "Society: Don't Blame the Mothers." *Nature* 512, no. 7513: 131–32.

Roberts, Celia. 2003. "Drowning in a Sea of Estrogens: Sex Hormones, Sexual Reproduction and Sex." *Sexualities* 6, no. 2: 195–213.

Roberts, Celia. 2007. *Messengers of Sex: Hormones, Biomedicine and Feminism.* Cambridge: Cambridge University Press.

Roberts, Celia. 2015. *Puberty in Crisis: The Sociology of Early Sexual Development.* Cambridge: Cambridge University Press.

Roberts, Celia. 2017. "Endomaterialities: The Toxic Politics of Hormones, Sex/Gender and Reproduction." In *Gender: Matter*, edited by Stacy Alaimo, 297–311. New York: Macmillan Interdisciplinary Handbooks. http://eprints.lancs.ac.uk/78117.

Roberts, Dorothy. 2011. *Fatal Invention: How Science, Politics, and Big Business Re-create Race in the Twenty-First Century.* New York: New Press/ORIM.

Roberts, Elizabeth F. S. 2017. "What Gets Inside: Violent Entanglements and Toxic Boundaries in Mexico City." *Cultural Anthropology* 32, no. 4: 592–619.

Roberts, Jody A. 2010. "Reflections of an Unrepentant Plastiphobe: Plasticity and the STS Life." *Science as Culture* 19, no. 1: 101–20.

Rofel, Lisa. 2007. *Desiring China: Experiments in Neoliberalism, Sexuality, and Public Culture.* Durham, NC: Duke University Press.

Rogaski, Ruth. 2004. *Hygienic Modernity: Meanings of Health and Disease in Treaty-Port China.* Berkeley: University of California Press.

Rogaski, Ruth. 2019. "Air/Qi Connections and China's Smog Crisis: Notes from the History of Science." *Cross-Currents: East Asian History and Culture Review* 8, no. 1: 165–94.

Roitman, Janet. 2013. *Anti-crisis.* Durham, NC: Duke University Press.

Rojas, Carlos, and Ralph A. Litzinger. 2016. *Ghost Protocol: Development and Displacement in Global China.* Durham, NC: Duke University Press.

Rose, Deborah Bird. 2017. "Shimmer: When All You Love Is Being Trashed." In *Arts of Living on Damaged Planet: Ghosts and Monsters of the Anthropocene*, edited by Anna Lowenhaupt Tsing, Nils Bubandt, Elaine Gan, and Heather Anne Swanson, G51–G64. Minneapolis: University of Minnesota Press.

Sahlins, Marshall. 2013. *What Kinship Is—and Is Not.* Chicago: University of Chicago Press.

Saldaña-Tejeda, Abril. 2018. "Mitochondrial Mothers of a Fat Nation: Race, Gender and Epigenetics in Obesity Research on Mexican Mestizos." *BioSocieties* 13, no.2: 434–52.

Saldaña-Tejeda, Abril, and Peter Wade. 2019. "Eugenics, Epigenetics, and Obesity Predisposition among Mexican Mestizos." *Medical Anthropology* 38, no. 8: 664–79.

Sandelowski, Margarete, and Sheryl De Lacey. 2002. "The Uses of a 'Disease': Infertility as Rhetorical Vehicle." In *Infertility around the Globe: New Thinking on Childlessness, Gender and Reproductive Technology*, edited by Marcia C. Inhorn and Frank Van Balen, 33–51. Berkeley: University of California Press.

Savage, Jessica H., Allison J. Kaeding, Elizabeth C. Matsui, and Robert A. Wood. 2010. "The Natural History of Soy Allergy." *Journal of Allergy and Clinical Immunology* 125, no. 3: 683–86.

Scheper-Hughes, Nancy, and Margaret M. Lock. 1987. "The Mindful Body: A Prolegomenon to Future Work in Medical Anthropology." *Medical Anthropology Quarterly* 1, no. 1: 6–41.
Schneider, David M. 1984. *A Critique of the Study of Kinship*. Ann Arbor, MI: University of Michigan Press.
Schneider, Laurence A. 1989. "Learning from Russia: Lysenkoism and the Fate of Genetics in China, 1950–1986." In *Science and Technology in Post-Mao China*, edited by Denis Fred Simon and Merle Goldman, 45–65. Cambridge, MA: Harvard University Press.
Schug, Thaddeus T., Anne F. Johnson, Linda S. Birnbaum, Theo Colborn, Louis J. Guillette, David P. Crews, Terry Collins, et al. 2016. "Minireview: Endocrine Disruptors: Past Lessons and Future Directions." *Molecular Endocrinology* 30, no. 8: 833–47. https://doi.org/10.1210/me.2016-1096.
Sellers, Christopher C. 1997. *Hazards of the Job: From Industrial Disease to Environmental Health Science*. Chapel Hill: University of North Carolina Press.
Shapiro, Nicholas. 2015. "Attuning to the Chemosphere: Domestic Formaldehyde, Bodily Reasoning, and the Chemical Sublime." *Cultural Anthropology* 30, no. 3: 368–93.
Shapiro, Nicholas, and Eben Kirksey. 2017. "Chemo-ethnography: An Introduction." *Cultural Anthropology* 32, no. 4: 481–93.
Sharp, Lesley A. 2018. *Animal Ethos: The Morality of Human-Animal Encounters in Experimental Lab Science*. Berkeley: University of California Press.
Shostak, Sara. 2005. "The Emergence of Toxicogenomics: A Case Study of Molecularization." *Social Studies of Science* 35, no. 3: 367–403. https://doi.org/10.1177/0306312705049882.
Shostak, Sara. 2013. *Exposed Science: Genes, the Environment, and the Politics of Population Health*. Berkeley: University of California Press.
Sivin, Nathan. 1987. *Traditional Medicine in Contemporary China*. Ann Arbor: University of Michigan Press.
Skinner, Michael K. 2010 "Metabolic Disorders: Fathers' Nutritional Legacy." *Nature* 467, no. 7318: 922–23.
Sleeboom-Faulkner, Margaret. 2013. "Latent Science Collaboration: Strategies of Bioethical Capacity Building in Mainland China's Stem Cell World." *BioSocieties* 8, no. 1: 7–24.
Society of Toxicology. 2006. "Animals in Research: The Importance of Animals in the Science of Toxicology." www.toxicology.org/pubs/docs/air/AIR_Final.pdf.
Solomon, Harris. 2016. *Metabolic Living: Food, Fat, and the Absorption of Illness in India*. Durham, NC: Duke University Press.
Song, Priscilla. 2017. *Biomedical Odysseys: Fetal Cell Experiments from Cyberspace to China*. Princeton, NJ: Princeton University Press.
"'Sperm Crisis' Approaching China." 2005. *Sina News*, April 27.
Steinhardt, H. Christoph, and Fengshi Wu. 2016. "In the Name of the Public: Environmental Protest and the Changing Landscape of Popular Contention in China." *China Journal* 75 (January): 61–82.
Stern, Rachel E. 2013. *Environmental Litigation in China: A Study in Political Ambivalence*. Cambridge: Cambridge University Press.

Strathern, Marilyn. 1991. "Partners and Consumers: Making Relations Visible." *New Literary History* 22, no. 3: 581–601.

Strathern, Marilyn. 2004. *Partial Connections*. Walnut Creek, CA: AltaMira.

Strathern, Marilyn. 2005. *Kinship, Law and the Unexpected: Relatives Are Always a Surprise*. New York: Cambridge University Press.

Strathern, Marilyn. 2013. "Environments Within: An Ethnographic Commentary on Scale." *HAU: Masterclass Series* 2: 207–39.

Sui, Suli, and Margaret Sleeboom-Faulkner. 2010. "Choosing Offspring: Prenatal Genetic Testing for Thalassaemia and the Production of a 'Saviour Sibling' in China." *Culture, Health & Sexuality* 12, no. 2: 167–75.

Sun, Wanning. 2015. "Cultivating Self-Health Subjects: Yangsheng and Biocitizenship in Urban China." *Citizenship Studies* 19, nos. 3–4: 285–98.

TallBear, Kim. 2015. "An Indigenous Reflection on Working beyond the Human/Not Human." *GLQ: A Journal of Lesbian and Gay Studies* 21, nos. 2–3: 230–35.

TallBear, Kim. 2018. "Making Love and Relations beyond Settler Sex and Family." In *Making Kin Not Population: Reconceiving Generations*, edited by Adele Clarke and Donna J. Haraway, 145–64. Chicago: Prickly Paradigm.

Thompson, Charis. 2005. *Making Parents: The Ontological Choreography of Reproductive Technologies*. Cambridge, MA: MIT Press.

Thompson, Charis. 2013. *Good Science: The Ethical Choreography of Stem Cell Research*. Cambridge, MA: MIT Press.

Todd, Zoe. 2016. "An Indigenous Feminist's Take on the Ontological Turn: 'Ontology' Is Just Another Word for Colonialism." *Journal of Historical Sociology* 29, no. 1: 4–22.

Todd, Zoe. 2017. "Fish, Kin and Hope: Tending to Water Violations in Amiskwaciwâskahikan and Treaty Six Territory." *Afterall: A Journal of Art, Context and Enquiry* 43, no. 1: 102–7.

Tousignant, Noémi. 2018. *Edges of Exposure: Toxicology and the Problem of Capacity in Postcolonial Senegal*. Durham, NC: Duke University Press.

"Toxicology." 2019. National Institute of Environmental Health Sciences, May 23. www.niehs.nih.gov/health/topics/science/toxicology/index.cfm.

Tracy, Megan. 2010. "The Mutability of Melamine: A Transductive Account of a Scandal." *Anthropology Today* 26, no. 6: 4–8.

Traweek, Sharon. 1992. *Beamtimes and Lifetimes: The World of High Energy Physicists*. Cambridge, MA: Harvard University Press.

Tsing, Anna Lowenhaupt. 2015. *The Mushroom at the End of the World: On the Possibility of Life in Capitalist Ruins*. Princeton, NJ: Princeton University Press.

Tu, Weiming. 2001. "The Ecological Turn in New Confucian Humanism: Implications for China and the World." *Daedalus* 130, no. 4: 243–64.

US National Institute of Health. 2013. "Hirschsprung Disease." Genetics Home Reference, December 10. http://ghr.nlm.nih.gov/condition/hirschsprung-disease.

Valdez, Natali. 2018. "The Redistribution of Reproductive Responsibility: On the Epigenetics of 'Environment' in Prenatal Interventions." *Medical Anthropology Quarterly* 32, no. 3: 425–42.

Valdez, Natali. 2021. *Weighing the Future: Race, Science, and Pregnancy Trials in the Postgenomic Era*. Berkeley: University of California Press.

Vogel, Sarah A. 2013. *Is It Safe? BPA and the Struggle to Define the Safety of Chemicals.* Berkeley: University of California Press.

Waggoner, Miranda R., and Tobias Uller. 2015. "Epigenetic Determinism in Science and Society." *New Genetics and Society* 34, no. 2: 177–95.

Wahlberg, Ayo. 2018a. "Exposed Biologies and the Banking of Reproductive Vitality in China." *Science, Technology and Society* 23, no. 2: 307–23. https://doi.org/10.1177/0971721818762895.

Wahlberg, Ayo. 2018b. *Good Quality: The Routinization of Sperm Banking in China.* Berkeley: University of California Press.

Walker, Brett. 2011. *Toxic Archipelago: A History of Industrial Disease in Japan.* Seattle: University of Washington Press.

Wanderer, Emily Mannix. 2017. "Bioseguridad in Mexico: Pursuing Security between Local and Global Biologies." *Medical Anthropology Quarterly* 31, no. 3: 315–31.

Wang, Bo. 2019. "Sacred Trash and Personhood: Living in Daily Waste-Management Infrastructures in the Eastern Himalayas." *Cross-Currents: East Asian History and Culture Review* 8, no. 1: 224–48.

Wang, Chuantao. 2010. "Humans Harm Yangtze River, Yangtze River Harms Fish" [*Ren Haile Changjiang Changjiang Haile Yu*]. *Henan Shangbao (Henan Business Daily),* August 31.

Warin, Megan, Emma Kowal, and Maurizio Meloni. 2020. "Indigenous Knowledge in a Postgenomic Landscape: The Politics of Epigenetic Hope and Reparation in Australia." *Science, Technology, & Human Values* 45, no. 1: 87–111.

Warin, Megan, Tanya Zivkovic, Vivienne Moore, and Michael Davies. 2012. "Mothers as Smoking Guns: Fetal Overnutrition and the Reproduction of Obesity." *Feminism & Psychology* 22, no. 3: 360–75.

Wentzell, Emily. 2019. "Treating 'Collective Biologies' through Men's HPV Research in Mexico." *Medicine Anthropology Theory* 6, no. 2: 49–71.

Wiley, Andrea S. 2015. *Re-imagining Milk: Cultural and Biological Perspectives.* New York: Routledge.

World Health Organization. 2013. "State of the Science of Endocrine Disrupting Chemicals—2012." www.unep.org/resources/report/state-science-endocrine-disrupting-chemicals.

World Health Organization Task Force on Methods for the Regulation of Male Fertility. 1996. "Contraceptive Efficacy of Testosterone-Induced Azoospermia and Oligozoospermia in Normal Men." *Fertility and Sterility* 65, no. 4: 821–29.

Wu, Jie. 2010. "Who Will Become the Next Hormone Containing 'Wild' Fish?" [*Shei Shi Xia Yi Tiao Han Jisu de "Yesheng" Yu?*]. *Commercial Times (Gongshang Shibao),* August 31.

Wylie, Sara. 2012. "Hormone Mimics and Their Promise of Significant Otherness." *Science as Culture* 21, no. 1: 49–76.

Xu, Qiong. 2010. "Yangtze River Poison Fish Not Only a Food Safety Concern" [*Chang Jiang Du Yu Buzhi Youguan Shipin Anquan)*]. *Chengdu Economic Daily (Chengdu Shangbao),* September 2.

Yan, Yunxiang. 2010. *The Individualization of Chinese Society.* Oxford: Bloomsbury Academic.

Yan, Yunxiang. 2012. "Food Safety and Social Risk in Contemporary China." *Journal of Asian Studies* 71, no. 3: 705–29.

Yang, Mayfair Mei-hui. 1994. *Gifts, Favors, and Banquets: The Art of Social Relationships in China*. Ithaca, NY: Cornell University Press.

Yang, Pu, Chongshan Zhong, and Lingdan Ge. 2010. "Experts Discuss Yangtze River Carp, Catfish Show 'Environmental Hormones'—Pay Attention, but No Need to Panic over Fish" [*Zhuanjia Tan Changjiang Liyu Nianyu Deng Cechu "Huanjing Jisu"—Ying Yu Zhongshi Dan Bubi Tan Yu Se Bian*]. *Xinhua Daily (Xinhua Ribao)*, September 1.

Yapp, Hentyle. 2021. *Minor China: Method, Materialisms and the Aesthetic*. Durham, NC: Duke University Press.

Yates-Doerr, Emily. 2020. "Reworking the Social Determinants of Health: Responding to Material-Semiotic Indeterminacy in Public Health Interventions." *Medical Anthropology Quarterly* 34, no. 3: 378–97.

Yeh, Wen-hsin. 2008. *Shanghai Splendor: Economic Sentiments and the Making of Modern China, 1843–1949*. Berkeley: University of California Press.

Zee, Jerry C. 2020. "Machine Sky: Social and Terrestrial Engineering in a Chinese Weather System." *American Anthropologist* 122, no. 1: 9–20.

Zelko, Frank S. 2013. *Make It a Green Peace! The Rise of Countercultural Environmentalism*. New York: Oxford University Press.

Zhan, Mei. 2011. "Worlding Oneness: Daoism, Heidegger, and Possibilities for Treating the Human." *Social Text* 29, no. 4: 107–28.

Zhang, Amy. 2020. "Circularity and Enclosures: Metabolizing Waste with the Black Soldier Fly." *Cultural Anthropology* 35, no. 1: 74–103.

Zhang, Chi. 2013. "Has Environmental Pollution Become an Important Factor of Infertility?" [*Huanjing Wuran Cheng Buyunbuyu Zhongyao Yinsu?*]. *Modern Express (Xiandai Kuaibao)*, December 23.

Zhang, Everett Yuehong. 2015. *The Impotence Epidemic: Men's Medicine and Sexual Desire in Contemporary China*. Durham, NC: Duke University Press.

Zhang, Everett, Arthur Kleinman, and Weiming Tu. 2010. "The Truth about the Death Toll of the Great Leap Famine in Sichuan: An Analysis of Maoist Sovereignty." In *Governance of Life in Chinese Moral Experience*, edited by Everett Zhang, Arthur Kleinman, and Weiming Tu, 78–96. London: Routledge.

Zhang, Joy Yueyue. 2012. *The Cosmopolitanization of Science: Stem Cell Governance in China*. New York: Palgrave Macmillan.

Zhang, Joy Y., and Michael Barr. 2013. *Green Politics in China: Environmental Governance and State-Society Relations*. New York: Pluto.

Zhang, Wenjun, and Shengyun Sun. 2015. "Negotiating the Theory of 'Shen Is the Essence of Xiantian, Pi Is the Essence of Houtian' in Chinese Medicine, from the Perspective of Epigenetics" [*Cong biaoguanyichuanxue tantao zhongyi "shen wei xiantian zhi ben, pi wei houtian zhi ben" lilun*]. *Journal of Guangzhou University of Traditional Chinese Medicine* (*Guangzhou Zhongyiyao daxue xuebao)* 32, no. 3: 559–65.

Zhou, Minghua. 2010. "Fish in Yangtze River 'Poisoned'"? [*Chang Jiang Li Yu Er You "Du?"*]. *Sichuan Daily (Sichuan Ribao)*, September 2.

Zhu, Jianfeng. 2013. "Projecting Potentiality: Understanding Maternal Serum Screening in Contemporary China." *Current Anthropology* 54, no. S7: S36–S44.

Zhu, Jianfeng, and Dong Dong. 2018. "From Quality Control to Informed Choice: Understanding 'Good Births' and Prenatal Genetic Testing in Contemporary Urban China." In *Routledge Handbook of Genomics, Health and Society*, 2nd ed., edited by Sahra Gibbon, Barbara Prainsack, Stephen Hilgartner, and Janelle Lamoreaux, 47–54. Abingdon: Routledge.

INDEX